Permaculture Volume 2

AF711389

Selected Articles from the Permaculture Association of Western Australia

Dr Ross Mars (Supervising Editor)

Permaculture Volume 2 -The Best of PAWA

Selected Articles from the Newsletters of the Permaculture Association of Western Australia

Ross Mars (Supervising Editor)

Acknowledgements

A sincere thank you goes to the following people who have made valuable contributions to this book:

Editorial team: Darrell Beck, Jennifer Payritz, Anna Agterhuis, Sue Hoolahan

Foreword: Max Lindegger

Proofreading: Pat Scott, Julie Woodman, Margaret Sampey

Artwork: Darrell Beck

Additional artwork: Simone Willis, Christopher Spurr, Gary Dorn, Patricia Dundas, Peter Hicks

Cartoons: Peter Robertson

Typesetting and layout: Ross Mars

Cover: Natalia Milianku, aged 10. "What permaculture means to me". Reference: Vol 16, number 1 (1994)

Every effort has been made to contact individual authors. However, this has not been completely successful. Enquiries about the contents of this publication can be directed to Candlelight Trust.

National Library of Australia

Cataloguing-in-publication data

Mars, Ross

Permaculture Volume 2: The Best of PAWA. Selected Articles from the Newsletters of the Permaculture Association of Western Australia

ISBN 0 9587626 2 7 (set)

ISBN 0 9587626 8 6 (v. 2)

1. Permaculture. 2. Organic gardening.

I. Mars, Ross. II. Permaculture Association of Western Australia.

III. Title: PAWA Newsletter.

635.0484

Forward

Max Lindegger

I've had the good fortune of having regular contact with PAWA for many years. We in the east have always been a little envious about the success of the group (size does matter!) and its quality newsletter.

Visits in recent years have shown that the enthusiasm has not been lost. The very successful Permaculture Conference in 1996 proved that permaculture has indeed matured and expanded well beyond gardens.

The editorial team has assembled a varied and well-chosen range of articles of interest, not just for the beginner but also for the experienced practitioner.

The publication is proof that permaculture is relevant, possibly more important in the twenty-first century than twenty years ago.

PAWA's newsletter is highly regarded throughout the world. It is appreciated that these articles, written by practitioners with hands-on experience, are available to a wide readership.

Congratulations to PAWA for promoting sustainability in an easy-to-read format and producing a fine reference book of timeless ideas and innovative strategies - something for everyone.

Max Lindegger, Crystal Waters, Queensland.

Contents

Urban Permaculture

- Permaculture *Gerald Shepperson* Reference: Vol 3, number 3 — 6
- Permaculture gardening in the backyard - practice and theory *Ray Hart* Reference: Vol 2, number 1 — 7
- Urban permaculture design *Peter Cocks* Reference: Vol 7, number 4 — 9
- Starting a permaculture system on a suburban block *Gerald Shepperson* Reference: Vol 8, number 3 — 11
- FINCing about a community *Martin Anda* Reference: Vol 17, number 3 — 11

Designing for a Sustainable Future

- Thinking about your future *Gary Dorn* Reference: Vol 15, number 1 — 12
- Designing for catastrophe *Warwick Rowell* Reference: Vol 14, number 1 — 13
- Ethics in permaculture *Ross Mars and Robert Troeth* Reference: Vol 18, number 4 — 17

Useful Plants in Permaculture

- Bamboo - a most useful plant *Wendy Wilkins* Reference: Vol 16, number 6 — 20
- Choosing fruit trees *Margaret Leslie* Reference: Vol 15, number 2 — 22
- Bush tucker *Ross Mars* Reference: Vol 15, number 2 — 23
- Our future with hemp *Martin Ducker* Reference: Vol 17, number 5 — 26
- Capuli cherry *Pat Scott* Reference: Vol 16, number 6 — 27
- Moringa oleifera - drumstick tree *Yvonne Pantino* Reference: Vol 22, number 4 — 27
- Legume trees - permaculture forage *Ros Hart* Reference: Vol 6, number 3 — 28
- Acacias in permaculture *Bob Hammond* Reference: Vol 11, number 1 — 29
- The deadly difference - what's in a name? *Miles Durand* Reference: Vol 14, number 4 — 30

Vegetables

- Celebrity plant profiles - sweet potato and cape gooseberry *Pat Scott* Reference: Vol 17, number 1 — 32
- Easily grown leaf vegetables *Margaret Leslie* Reference: Vol 4, number 2 — 32
- Useful gardening hints - jerusalem artichokes and perennial silverbeet *Pat Scott* Reference: Vol 11, number 4 — 33
- Successful bean growing *Margaret Leslie* Reference: Vol 8, number 3 — 34
- Abundant surprises from permaculture *Heather Lamont* Reference: Vol 16, number 2 — 35
- Growing tips for vegetables *Margaret and Leo Leslie* Reference: Vol 13, 15, numbers 3 and 4 respectively — 36

Herbs

- A herb and salad picking bed for vibrant good health *Valerie Hopkins* Reference: Vol 10, number 4 — 38
- Wormwood *Pat Scott* Reference: Vol 16, number 6 — 39
- The family umbelliferae *Debbie Milianku* Reference: Vol 17, number 5 — 39

- Herb focus - the mint family *Ross Mars* Reference: Vol 22, number 3 40
- Pineapple sage *Pat Scott* Reference: Vol 17, number 1 41
- English dandelion - weed or treasure to behold? *Jan O'Keefe* Reference: Vol 21, number 4 42
- Wild herb walk *Sue-Anne McDonald* Reference: Vol 19, number 4 42

Soils

- Soil fertility *Halin Orion* Reference: Vol 23, number 1 43
- Soil life - buried treasure *Margaret Howe* Reference: Vol 22, number 6 48
- Farming for soil health *Robert Campbell* Reference: Vol 18, number 2 49
- The big nitrogen fix *Miles Durand* Reference: Vol 12, number 4 50
- From soil to plants *Christopher Spurr* Reference: Vol 20, numbers 1 and 2 52
- Much to do about mulching - a permaculture perspective *Miles Durand* Reference: Vol 14, number 5 54
- Mulching *Halin Orion* Reference: Vol 23, number 2 56
- On brushwood composting *Mel Van Der Elst* Reference: Vol 13, number 2 58

Pest Control

- A natural alliance between permaculture and integrated pest management *Sandra McDougall* Reference: Vol 18, number 6 60
- See the solution - not the problem! *Dora Byrne* Reference: Vol 20, number 5 63
- Integrated fruitfly management *David Piggott* Reference: Vol 19, number 1 65
- A slater problem? or a quail deficiency? *Yvonne Pantino* Reference: Vol 21, number 5 70
- Handy hints for pest control Reference: *various* 71

Appropriate Building Materials

- An owner built passive solar house *Doug McLennon* Reference: Vol 8, number 4 73
- Retrofitting house *John Barker* Reference: Vol 2, number 4 74
- Mudbricks or rammed earth? *Michael George* Reference: Vol 14, number 6 79
- Building with rammed earth *Steve Dobson* Reference: Vol 14, 19, 21, numbers 6, 1, and 5 respectively 80
- Organic solar architecture - the synthesis of strawbale construction and renewable energy *Gary Dorn* Reference: Vol 18, number 4 83
- An integrated solar house *Ross Mars* Reference: Vol 18, number 5 86
- Solar architecture revisited *Gary Palma* Reference: Vol 22, number 6 89
- Strawbale building *Ed Taylor* Reference: Vol 20, numbers 4 and 5 90

Use of Renewable Energy

- Building your own solar hot water system *Frank Legge* Reference: Vol 1, number 3 92
- The production of methane gas *David Coleman* Reference: Vol 9, number 4 94
- Simple food drying *Dave (Sunny) Miller* Reference: Vol 23, number 1 95

Appropriate Technology and Techniques

- Alternative technology - large scale chicken tractor *Dave Edwards* Reference: Vol 2, number 1 96
- Notes on fruit drying *Margaret Leslie* Reference: Vol 11, number 4 98
- Irrigation *Gerald Shepperson* Reference: Vol 9, number 1 98
- Raised beds for Vegetables *Gerald Shepperson* Reference: Vol 7, number 3 99
- Electric fences *Peter Linehan* Reference: Vol 3, number 3 100
- Sulphuring fruit before drying *Rosalind Rennie* Reference: Vol 5, number 2 101
- Building a small grain mill *Matt Bruekers* Reference: Vol 13, number 3 102
- Plant support systems *Gerald Shepperson* Reference: Vol 10, number 1 103
- General handy hints Reference: *various* 105

Waste Management

- To sludge or not to sludge? *Floris Gout* Reference: Vol 16, number 3 106
- Distributing greywater *Bob Hammond* Reference: Vol 11, number 4 109
- Water conservation *Peter Mason* Reference: Vol 11, number 4 109
- Greywater reuse at Candlelight Farm *Ross Mars* Reference: Vol 17, number 1 110
- The Clivus Multrum composting toilet *Miles Durand* Reference: Vol 14, 17, numbers 1 and 1 respectively 112

- The Dowmus composting toilet *Dean Cameron* Reference: Vol 17, number 1 114
- Urine in the garden or 'wee will fertilise' *Alanna Moore* Reference: Vol 16, number 1 114

Permaculture Education in Schools

- Information for schools wishing to implement permaculture-style gardens *Ross Mars* Reference: Vol 17, number 5 115
- School gardens - grounds for learning *Jim Smith* Reference: Vol 16, number 4 117
- The outside classroom - permaculture in schools *Ross Mars* Reference: Vol 22, number 5 118
- Permaculture education for children *Colin Bygraves* Reference: Vol 14, number 2 121
- Teach your children well (parts 1 and 2) *Nathan Floyd* Reference: Vol 20, 21, numbers 6 and 1 respectively 122
- Permaculture Kalumburu style *Margot-Anne Barefoot* Reference: Vol 19, 20, numbers 1 and 1 respectively 124

Propagation

- Successful method of raising trees from hard-shelled seeds *Roger Morgan* Reference: Vol 6, number 3 127
- Growing native trees from seed *Julie Woodman* Reference: Vol 10, number 4 128
- How to take cuttings *Yvonne Pantino* Reference: Vol 22, number 5 128
- Vegetable seed saving *Shirley Rowlands* Reference: Vol 13, number 3 130
- How to store seed *Ross Mars* Reference: Vol 23, number 6 133
- Handy hints for propagation Reference: *various* 135

Special addition : a poem

- What a wasteful lot or a permy's paradise *Jeremy Parker* Reference: Vol 19, number 4 137

Other Publications by Candlelight Trust

All of the following are available from Candlelight Trust, 4110 Phillips Rd, Mundaring. Western Australia. 6073. www.cfpermaculture.com.au

Books (printed)

1. Mars, Ross and Mars, Jenny. (1998). *Getting Started in Permaculture.* ISBN 0-646-20090-9. 100pp. (Revised from 1994 ed.).
2. Mars, Ross. (1996). *The Basics of Permaculture Design.* ISBN 0-9587626-1-9. 170pp.

e-Books

1. Mars, Ross and Willis, Robert. (Eds). (1996). *Permaculture Volume One: Best of PAWA - selected articles from the Newsletters of the Permaculture Association of Western Australia.* ISBN 0-9587626-3-5.

Videos/DVD's

1. *Passive Solar Design of Buildings.* ISBN 0-9587626-4-3. 23 mins. Narration: Peter Holland.
2. *RAPS - Remote Area Power Supply.* ISBN 0-9587626-6-X. 28 mins. Narration: Verity James.

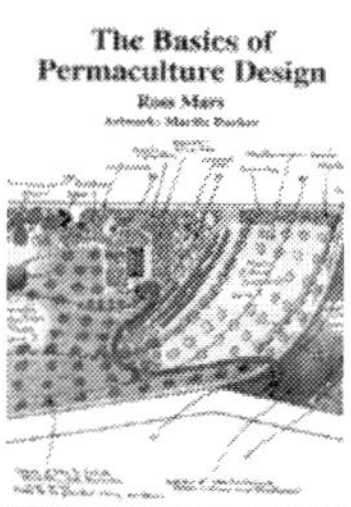

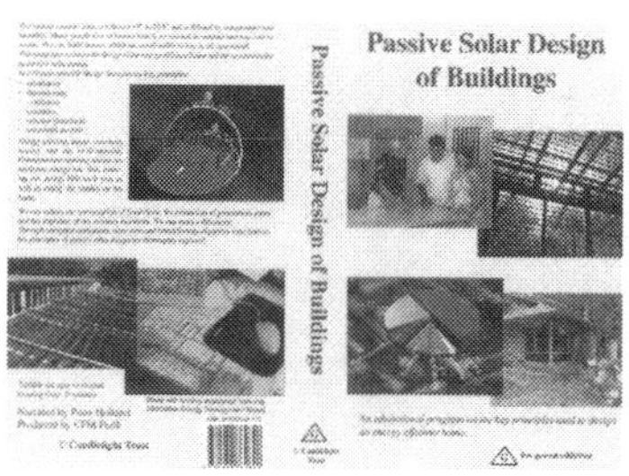

Permaculture

The general meaning of the word 'Permaculture' is permanent agriculture: a system which would stand a fairly good chance of survival without too much help from us. There are no fixed rules. We tore up the rule book when we decided to change from conventional agriculture methods. Over the last few decades, at least in the Western world where people can afford it, people have come to rely on chemicals, both in the form of fertilisers and poisonous sprays. The same goes for our dwellings where we have been in a position to install an oil or gas heater and pay for the oil or gas quite easily. It is obvious that this trend is coming to an end due to the problems they cause and their price. An answer can be found with a Permaculture system, which will include your house and block together with your gardens.

Your system will be unique, according to differences in:

* The direction in which your house faces.
* The existing trees and shrubs already on the block.
* What fences you have.
* What is built or growing on the adjacent blocks.
* What your local government regulations might be.

When taking these five items into consideration we can now appreciate that copying a successful system created in one suburb or street will not guarantee success in different surroundings.

Around the house

It is a good idea to start with the house as we can make various improvements that are permanently advantageous to our comfort and help us economically. We can begin by improving ventilation for the hot weather and installing insulation for the cold weather, bearing in mind that insulation to keep heat in in winter also keeps heat out in summer. Around the house, where possible, we can grow deciduous trees, which will keep the sun out from the east and the west. In the north, the trees will shade the ground to the extent that you will not get heat bouncing up through the windows. A small addition on the roof overhang, making it one metre wide approximately, will keep the sun completely off four meters of north-facing wall. On the west side of the house, a deciduous vine is very good as this insulates in summer but allows the wall to heat up in winter. At the back door, which is handy to the kitchen in most cases, many of us like a herb garden, even one planted in hanging baskets. I have my herb hanging baskets under a Cape Lilac tree, with paving beneath which makes a very pleasant barbecue area.

In the garden

In the rest of the garden, we need a certain amount of shade, which may already be there on established blocks. We may also want a windbreak against the hot winds off the land. If we live near the ocean, we would resist the salt-laden winds with salt-tolerant windbreaks. Don't cut anything down without studying to what advantage it could be put. Even then, give it second thought. If you take a shrub or a tree out, you will lose it in three ways: shade, windbreak and mulch. Most trees and shrubs will survive a very severe cutting back, even to the butt. You can cut all this material up for mulch, which is suitable to put around and between your fruit trees.

Chickens and ducks

The next two items depend on your Council by-laws - chickens and ducks. Chickens for eggs, meat, fertiliser and natural tractor; ducks for eggs, meat and a natural predator for snails. My own idea for a chicken tractor in a back garden is to use a galvanised garden shed, approximately 3 m x 2 m with a front and rear door. Build a run at both the front and back approximately 7 m x 3 m allowing about twelve chickens to work one run for about six months. Vegetables can be grown in the first run whilst the chickens move on to the second. The situation can be changed to suit the crops, ie. winter crops starting in March or April comprising potatoes, carrots, onions and garlic, and then the summer crops of beans, tomatoes, cucumbers and lettuce.

You will find that the chickens will dig and fertilise the soil to a depth of about 20 cm. The floor under the shed must have an impervious floor. The droppings and litter from the chicken house are best placed in the compost bin to avoid breeding the stickfast flea. With a shed the size I have mentioned, there is room to keep this year's chickens until you eat last year's.

Your ducks can roam the whole garden, although you may need to put low fencing around small plants to prevent them from being eaten. When you get a few young ducks, you will most probably have to confine them to an enclosure. Ducks are well worthwhile for cleaning up your snails and for duck and plum sauce afterwards, plus a few eggs to put in your cakes.

Fruit, vegetables and bees

We can now turn to the areas we have left, bearing in mind our likes and dislikes. We chose to plant grapes and passionfruit in a sunny spot, also citrus trees, plums, apples, fig, pomegranate and so on. We need to plant trees about 4 metres apart as we need to get in between to pick the fruit. Things like sweet potatoes, New Zealand spinach, rhubarb and perennial leeks can be grown in almost any corner. In the Permaculture system, we put plenty of compost on all these areas and a good mulch on top of at least 100 mm of straw. If there are any problems with the straw not breaking down due to a lack of worms, just sprinkle chicken droppings on top of the straw, before adding any more, and the worm activity will increase immensely within ten days.

One other activity you need in your system is a large number of visits from bees. Encourage bee visits by planting ornamentals all over the garden. These need not be very large plants as they only need to be attractive to the bees without interfering with anything else.

Following a system like this is enormously labour saving, with a great increase in output compared with input.

Permaculture Gardening in the Backyard - Practice and Theory

Before I describe what we have been doing I will list the basic information that is important. We also made it simpler by assuming that we were not prepared to do any major alterations to the house (partly because we didn't want to spend money and because we didn't intend to be in the house for long).

1. In winter it rains a lot, the weather is mild to warm, frosts are not really a problem and sunlight can be limited.

2. In summer there is virtually no effective rainfall, it is excessively hot and there is super-abundant sunlight.

3. The soil is mostly sand devoid of organic matter, with a low water-holding capacity and rapid water loss to the air by wind and/or sun.

4. Wind is a problem. It can be destructive in winter as a westerly and the summer problem wind is the hot and drying easterly.

5. Plants with a high chill requirement may not fruit successfully on the coastal plain, but they may in the hills.

6. The rainfall and temperatures vary substantially from west to east. It is noticeably wetter, colder in winter and hotter in summer as you go eastward away from the coast, and this is accentuated as you go up to the hills.

7. The sun is low in winter casting long shadows, but is virtually overhead in summer.

With these considerations in mind we have been evolving a workable system. The basic themes and design principles we have used are as follows:

1. Treat with a considerable kilogram of salt everything the books tell you.

2. Think in terms of using, not fighting, the obvious features of this climate and soil. In summer this climate is perfect for figs, grapes, olives, mulberries, loquats, cape gooseberries, mediterranean herbs (like sage, rosemary, thymes, basil and yarrow), lemongrass, all the melons, corn, eggplants and many beans. In winter, it is a very suitable place to grow most of the common vegetables, some herbs, like borage, and grains.

3. If some plant gives you trouble give it a miss. If your lettuces drive you crazy by bolting in spring and summer don't grow them, find a new variety or better still throw them away and grow a perennial summer-loving crop like New Zealand Spinach for salad greens. (What's more you don't have to keep on buying seed of the spinach, you can divide it and give it away all the time.)

4. Think in terms of perennial plants. These really do have lots of advantages.

(a) They tolerate the summer drought vastly better than the annuals. A New Zealand Spinach will literally leave a lettuce for dead at 40°C without copious watering.

(b) They produce very good yields with little care. A grapevine takes about an hour to prune, minimal watering and will produce wheelbarrow loads of grapes.

(c) Many produce fruit in summer, which is the ideal time for eating fruit. Fresh loquats are fine in the spring, followed by mulberry pies, then figs for lunch in summer and cold grapes from the fridge.

(d) They produce shade, shelter, flowers, cool air, soil nutrients in leaf fall, and are noise suppressors and air cleaners.

5. Don't feel limited by space restrictions. You can fit a grapevine in anywhere. There are numerous shrub to small tree-sized plants available such as plums, persimmons, figs, citrus, hazelnut, feijoas, all the guavas, and a lot more you probably haven't even heard of. Once the jungle gets going we can bring in understorey plants like tamarillos easily.

6. Get into compost for the soil. We use basically any high carbon material (grass clippings, straw, weeds, garden rubbish) put through the poultry pen where it is enriched with manure and scraps, variously compacted and shredded, and grows soil bugs for the birds to eat. We also sometimes make a conventional compost pile, but the poultry have largely superseded that. We got rid of the compost tumbler. We also add seaweed to the compost, primarily for trace elements and minerals.

Compost increases the water holding capacity of the soil, makes nutrients available to plants and gives soil-borne diseases and pests (like nematodes and fungi) a hard time.

7. Get into mulching. This really does work and it needs to be thick. We use straw and grass clippings, and sometimes newspaper if there is an isolated plant requiring urgent help or a weed problem. Mulch does various things - it stops the sun and wind from heating up and drying out the soil, allows the soil life to come up and do their composting thing in peace, and engulfs weeds. But, it can also harbour pests and diseases, and stop light rain or watering from reaching the ground.

There are a lot of experiments going on now with mulches and their good and bad points. Bringing in weed seeds is a chronic problem. To cope with this we have designed a solar-powered weed killer. This complex piece of high technology consists of a discarded window from the dump. You lay this on the germinating weeds and the sun steams them to death. Imagine if the weeds tried to germinate inside the solar hot water system. Very neat. Most weeds will germinate over a short time in the autumn, but there are some tough summer ones like crabgrass and couch.

The biggest problem with mulch is that you cannot easily water a plant through the mulch. The mulch keeps water out as well as in so you must water the plant not the mulch. This means you don't use sprinklers. Use a trickle system or direct hose watering. Trickle is undoubtedly superior for trees and any larger permanent plant.

We have devised a good system for watering small plants under thick mulch. We carefully contour the ground first so that the plants are actually growing in little trenches or in shallow ponds, with the mulch laid over the top of the whole lot. The water is applied to flood the area around the plants under the mulch.

For this same reason we let the mulch compost away from late summer so that the winter rains can reach the soil. The mulch is re-applied in October when the rain becomes sporadic and water loss rises rapidly. You can see this effect clearly under leafy trees in winter - they are extraordinarily efficient at stopping rain from reaching the soil. Another important point is that we add blood and bone or some other high nitrogen component to ensure that the decomposing mulch does not rob the soil. Most mulches are very high carbon.

One other benefit from mulching is that it largely eliminates compaction due to trampling, which is a serious problem in sand.

8. Consider trellising for summer vegetables and small trees. It is too hot here in summer but there is plenty of sunlight. A trellis cuts the heat and sun down, allowing the plants to transpire without wilting. This stops leaf scorch which is due to a failure of transpiration - when the water runs out the leaf cannot lose heat by evaporation and is destroyed by the rising temperature. (Wilted leaves don't grow either).

We use bamboo culms and palm fronds for the trellis which are easy to take off and store in winter. Of course the best trellis is a living roof of grapevines, chokos and so on but this takes time.

9. Consider abandoning vegetable growing in the summer. Mulch all the trees like crazy, water occasionally, have a rest from gardening and go on holiday. We have compromised and grow only a few annual summer-hardy annuals like sunflowers over things like herbs to protect them in the front garden.

10. Plan the position of your trees. How about evergreens on the south and in odd places to stop winds, and deciduous trees on the north to complement the trellis and on the east to stop the hot easterlies.

This means in winter the bare trees (like figs) will let in the morning sun when it is needed for the vegetables.

11. Don't forget creepers. These can be fitted in anywhere - passionfruit, banana passionfruit, chokos, grapevines, giant beans and so on.

For gardening in small spaces the basic rules are go up and diversify. In summer this is easy when there is plenty of light but be careful in winter. Summer shade from creepers can be very useful.

12. Incorporate poultry into the system. They are very productive and useful.

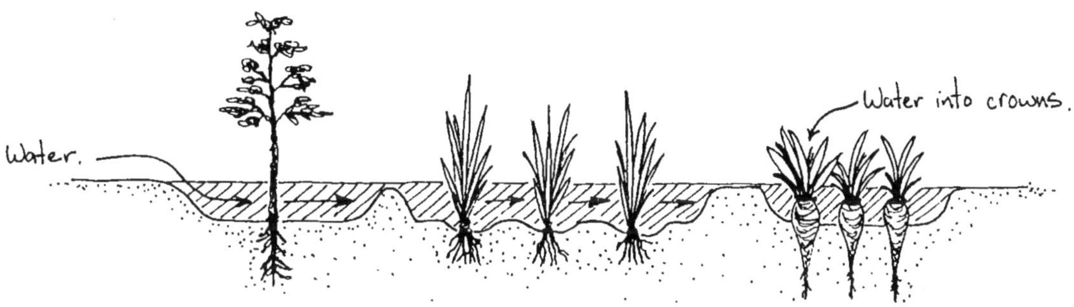

Urban Permaculture Design

The Permaculture home garden is a biological unit: we should 'aim' to produce all we need to avoid buying anything in. If starting from scratch, use sheet mulching to create at least 25 mm of good topsoil per year.

Composting and a worm farm are essential for good humus, which is what plants feed on. Never use straight manures (or artificial fertilisers) as they force-feed plants with soluble salts. Compost the manure first, into humus, or put it through your worm farm. Get the whole area alive with humus before you start creating beds - and work from the kitchen door outwards, using the zoning principles.

Zoning for an urban permaculture design

The zoning concept is basic to any permaculture design but should not be considered definitive. The place to start is the doorstep, and the zoning (or distance from the doorstep) is decided on two factors: a) the number of times you need to visit a plant, animal or structure; and b) the number of times the plant, animal or structure needs you to visit it.

For example, you may need a fresh lemon 60 -100 times a year, but the tree only needs you 6 - 12 times a year, making a total of 66 - 112 times. For an apple tree, where gathering is less, the total may be 15 times visited. The lemon tree is placed in zone 1, close to the doorstep, whereas the apple tree is well away in zone 2.

Zone 1 is the area surrounding the dwelling, which is used for the most intensive and controlled purposes and may include workshops, greenhouse, propagating frames and an intensive vegetable garden.

Plantings from doorstep outwards would include: culinary herbs at the kitchen door, lemon tree, clippings beds for small salad plants, path-side plucking and long-yield plants, often-used plants in narrow beds (often in raised beds), low yield plants (if there is room), windbreaks and weed barriers in, around and through your garden, and vine and trellis crops.

Zone 2 is the next area further away from the house, and would include: forage plants and structures for your small domestic animals, poultry, bees, a pond for your fish (though this could be in the greenhouse instead), a few orchard species, and various plants (mainly perennial herbs and trees) for essential oils, shelter, wildlife, medicinal value and production of income.

This is a suggested plan of course. All backyard permaculture plans must fit the individual situation.

Small animals can be kept if local ordinances allow (poultry, bees, quail, fish, guinea pigs).

Poultry should be placed for best advantage, for example, under fruit trees to control fruit fly, and yet near enough to allow frequent visits. Allow the use of a "chicken tractor" (a movable long box or cage with no floor containing some chooks) for weed and pest control, and to fertilise an area in preparation for replanting with vegetables. Choose the best breed for the situation. Light Sussex birds are good layers and good eating and are quiet enough for the backyard. Use a deep litter system for their main house floor. Plant their area with as many "poultry food" plants as your space permits.

Bees should be placed carefully to avoid stings so their flight path is not across anyone else's path.

General hints for suburban gardens:

- Reduce lead levels by planting screens of non-edible trees near roads.
- Take over the lawns in urban yards for fruit tree and vegetable production.
- Use dwarf varieties of fruit trees and espaliers and multigrafts, if block size or space is limited.
- Utilise small areas for planting, such as window boxes, porches, roofs (if flat).
- Put a glass house (hothouse) on the north side of the house to grow extra tropical fruit and vegetables, as well as to keep the house temperature warm in winter.
- Have a shadehouse or pergola on the south side of the house so that in the summer, when the hot air vents out of the north-facing glasshouse, it draws in the cool air from the shadehouse through the house to replace the lost hot air.
- Retrofit your suburban house to make it comfortable climatically and economical to run.

If you can't grow everything you need, grow what you can and barter or exchange the surplus with others doing the same, or organise like-minded people to plant in a local community garden.

Starting A Permaculture System on a Suburban Block

Anyone buying a property to start a permaculture system should go and look at it around midday. At this time your shadow will point away from you to the south. This gives you some idea about both shade and full sunlight areas for food production.

If your block is cleared it is much harder for the conventional person to steer away from a lot of the monoculture ideas which are so prevalent in today's western world. We must endeavour to get away from straight rows, huge areas of one crop and the dreaded digging in our type of climate. The most important thing is to develop the system towards its own self sufficiency.

Herb gardens should be as close as practical to the back door. You may have too much shade on an established block but only trim the trees as necessary if they serve as a windbreak. Might I also suggest planting mulberries as they are quick growing and are easily pruned to any shape. Chilean Willows are also a quick growing windbreak and provide good material to munch up and feed back on to the soil.

Just because a bush or tree has no edible use, don't despair. Hibiscus, oleander and many common shrubs found in the back garden feed the worms, and they in turn feed the vegetables that feed you.

Fruit trees are a very good investment in the backyard. Plant them 2 to 3 metres apart to save room. Prune them in later years so that you can just squeeze between them to harvest the fruit. If room is scarce, plant multi-grafted trees or two trees in one hole. Instead of pruning just cut some out when they get too big.

While your trees are small grow vegetables in between so that you can irrigate with a sprinkler without thinking what a waste every time you switch it on. Use a series of circular raised beds which are very convenient in small areas. Anything from 100 mm to 200 mm high will retain a good medium to grow vegetables in. When these are no longer viable move the retainers and level the good soil under your trees.

We have the snail problem so we have a couple of Khaki Campbell ducks. Snails in one end and eggs out of the other. In my opinion, ducks are better value than chooks in the system. Chooks can be added using two runs from one shed - chooks in one run whilst vegetables are grown in the other. Bantams give the highest return for the amount of food consumed. They can also be kept in a much smaller area. Whatever one chooses to keep, the manure they produce is most useful and helps towards the self-sufficient system we aim to achieve.

Just bear in mind that you feel like working if you sleep warm and dry in the winter and cool in the summer, and poultry will lay more eggs in similar circumstances. The more effort you put into your stock the more return you will reap. Good luck!

FINCing About a Community Garden

King William Park in South Fremantle is being transformed. The bare and barely-used patch of scraggly couch has been reshaped into dunes, terraces, swales and sheltered open areas by Fremantle Inner City Agriculture (FINCA) and other local residents.

The landscaping of the 800 square metre park is complete. The Community Garden project, which is on the corner of King William Street and Marine Terrace in South Fremantle, aims to provide a participatory process of education in permaculture, bush regeneration, landscape architecture and community arts.

The aim is a community-designed, community-made and community-maintained public garden. Limestone rubble retaining walls have been completed, the greywater reuse system has been installed and a ring mains for backup mains water irrigation has been buried.

The greywater reuse system will receive bathroom and laundry water from the two adjacent houses at higher level, which will be sub-surface irrigated to the terraces and the basin by gravity flow after having passed through a single septic tank.

The landscaping has been based on the concept plan which was prepared after workshops with the community and consultation with the project's permaculture designer, architect and restoration ecologist.

Dunes had been created along the Marine Terrace edge during the earthworks and these were planted out with local species.

Opportunities for the community art side of the project include: ceramic tiles in the paths and benches reflecting the area's cultural heritage, a bridge over the basin, creative children's play equipment, renovation of the brick shelter into a lookout with artistic rendering, decoration of the limestone rubble wall tops, murals on the surrounding walls... our imagination is limitless. It is hoped that the community arts process will also access residents who have not yet been involved in making the garden as well as opening us all to the creative delights of combining function and aesthetics.

Thinking About Your Future

There are a few ethics or philosophies in Permaculture systems which bring the client and designer together in an underlying common purpose or concern. These include considerations for low energy design, natural bio-degradable materials, ecologically sound conservatism, diversity and polyculture systems, lifestyle and concern for natural landscape.

All designers enjoy developing ideas and seeing them come into reality. There is a great deal of satisfaction in combining the above with all the conditions of site, structure, construction technologies, clients brief, by-laws and landscape into a creative harmonious solution.

I have found that the concern for most clients is "it must be functional and it must work". If people take the time to think at a higher level, the functional aspects become easier to design and in fact fit into place a lot easier. The difficulty then becomes in creating the beauty in the design.

I believe that true beauty comes from the spirit of a place, from a combination of feeling, sense of character, cultural aspirations, natural energies and landscape; where their unity touches our human consciousness.

The way I approach projects is by asking myself "how can I make this place really special and better?" I often ask my clients to undergo a short mental exercise where they can put their own ideas into some sort of 'big picture'. I begin by asking them to have a brainstorming session where they think about the project in three areas:

1. Physical - such as comfort, views, how do you want to feel.
2. Practical - functional aspects such as budget, workability, number of rooms, size and so on.
3. Spiritual - which includes personal ideas, thoughts on philosophy, aspirations and beauty.

You can try this yourself. Write whatever you think and as much as you can. List your ideas and let them flow; after a while they will come so quickly that you will need to really speed up your writing. It is important that you are not critical of your thoughts; don't judge as this limits your ideas. Do this exercise by yourself and then with others who are major occupants, contributors or users of your project.

The next step is to edit your ideas. This can be hard as you have to make choices and compromises. This is where you can begin to really take control and be responsible for your future, which I think is great. You have to think in terms of: important - must have, can't compromise; negotiable - size, location, form can change; desirable - would like to have but can be omitted if inappropriate.

This process may transform you or hopefully permit you to endow more trust onto your designer and give them greater freedom to interpret your brief and produce an innovative and rewarding design. One of the hardest things to do is to design for someone who doesn't know what they want.

The important thing about this in many ways is to think of your ideal situation, how you want to live or work, the quality of life that you want to live. Permaculture provides the principles and techniques so that we can create something special, a place of peace in which to rest and a place for our souls to dwell.

Designing For Catastrophe

Permaculture is not about sheet mulching or organic gardening. Permaculture is a sensitive process for designing sustainable systems for living. It is based on a collection of the new and old experience of huge numbers of people, all over the world. But it is more than the collection. It is about how each item in this collection, be it a technique, a species, or a piece of knowledge, can be placed in relation to another to make the system more resilient and better able to meet the needs of the people within it. Permaculture is about careful, prolonged and thoughtful observation, rather than prolonged and thoughtless labour. For example, permaculturists see the issue of global warming firstly from a systems perspective, and then from a practical, regional and local perspective.

Systems theory has contributed much to the way we see our world. It emphasises the inter-relationships between the elements of a system. Before a systems perspective was possible we looked at individual parts of the system from the perspective of say, physics, chemistry or biology. These frameworks have led to detailed but incomplete views of the world. Systems theory attempts to look at the features of the system as a whole and to identify which factors, internal and external, impact on its stability or vulnerability.

A systems view is often difficult. It is not easy to lift one's intellectual sights to the scales of time and space required. One of the major contributions recently has been that of the Gaia Theory, which advances the view 'the earth is alive' rather than 'life exists on earth'. This Gaian perspective requires a real stretch. Firstly, we need to think in geological time spans, where a million years is a terribly short time interval. We need to think about the movement of tectonic plates as short term, and the slow cycling of (comparatively) molten rocks underneath these plates as one of the key recycling systems of the planet - geology becomes planet physiology! Secondly, we need to trace the connections between bacteria and cells, and the planet-wide system. Thirdly, and probably the most difficult, we need to see ourselves as minor actors in this system, and subject to its influence.

Let's look at how systems behave. Systems are very stable - through a series of complex inter-relationships, they maintain a wide range of conditions at equilibrium, or between stable limits. For example, despite the Sun's radiation output increasing by 25% since the earth was formed, the atmosphere, the biosphere and the rocks and oceans of the earth have interacted in such a way as to preserve the earth's average temperature around 20° Celsius.

But when the stable point moves closer to the limits of the system (say 10° and 30° Celsius for this example), the oscillations around those (moving) average values become greater and greater. The stabilising mechanisms are not as sensitive, or as well timed, and so the system 'hunts' - it oscillates around the mean without settling to it. The commonest example we know is a sick child's fluctuating fever and chill. Another is the air-conditioned office that has poor placement of thermostats nearby the north facing panes of glass - it is a frustrating experience as you cycle from too cold to too hot.

With this fairly straightforward idea, we can start to explore the dense field of fact, interpretation, prediction, and criticism that makes up the burgeoning field of Climate Change. Present discussions about climate change and the greenhouse effect at least seem to agree on history, so let's start from there.

About 145,000 years ago the concentration of CO_2 in the atmosphere was 190 ppm. Over the next 5,000 years, it quite rapidly rose to 300 ppm. Then it slowly dropped back to 190 ppm 20,000 years ago. The CO_2 concentration has risen back to 275 ppm over the last 20,000 years. Looking at more recent history, from a level of 280 ppm in 1840, the concentration of CO_2 had risen to 345 ppm by 1990. So, in that last 150 years, the concentration has changed as much as it did over 5,000 years, way back, and already the concentration exceeds previously known levels by 20%.

It is also agreed that if today we could stop all CO_2 emissions, levels would continue to rise for about another 50 years; let's conservatively say to about 360 ppm. As well, it is agreed that CFCs and other industrial gases, such as methane, and nitrogen oxides, contribute to greenhouse effects much more potently that CO_2.

Yet the intense and expensive world-wide discussions about greenhouse effects are about what climate changes will occur if we double 1900 levels of CO_2 in the atmosphere - from 300 ppm to 600 ppm; or is it 350 ppm to 700 ppm by 2025? We already have enough data to judge that the system is reaching its limits.

What happens when systems reach the limit of their control mechanisms? We know that volcanoes and meteorites have had effects on climate that are much more devastating than the change we are talking about. One of the heartening things about the Gaian analysis of the Earth's life story is to discover that the balances can re-stabilise from these massive, unpredictable events.

This ability to re-stabilise is much more than balance. A system will not survive if the disturbing events consistently occur closer together in time than it takes the system to adapt to the last change. For example, people can have their mental and physical health substantially reduced by a number of small changes in their daily lives if they occur all at once, or fairly quickly one after another.

So a crucial question to ask is will the system have time to recover its stability, or are there other factors which might push it into catastrophe? This question is much more important than the detail of the exact amount of CO_2, and what causes it, and arguments about whether temperatures might rise 1° or 2°C. We must take a broader view.

Greenhouse effects are primarily the result of many industrial processes, many of which involve burning fossil fuels. Without going into the fine detail, we can see that increasingly rapid removal of tree cover is also pushing vegetative control systems of the planet to the limit. Any satellite photo of the globe now reveals the huge areas of pale yellow desert across North Africa, the Middle East, and right across southern Asia to the Chinese coast.

As recently as 3,000 years ago much of that area was under dense and complete forest systems. Just using our eyes shows us the impact increased reflectivity has on what was previously a more stable system, without further statistics about desertification, and the consequent drought and famine.

So far we have identified two system-wide impacts of increasing human population and our associated industrial and agricultural systems. At a more human scale, we see worldwide the impacts of urbanisation and pollution, and the social and health breakdowns that result. A systems perspective would also classify these as signs of system overload.

It is not being alarmist, but rather pragmatic, to suggest that we humans are placing the life system of the planet under threat. If the Gaian perspective is correct, then our threat is one that will be easily dealt with. One of the design features of any system is an ability to remove and isolate subsystems that go out of control, and threaten the whole system. The living earth has repeatedly and systematically eliminated threats to its existence - even if they are its own creatures.

Permaculture can be described as a contraction of both permanent agriculture and permanent culture. Its focus is designing, setting up and maintaining systems for living that are sustainable in the long term, and some of the major principles of permaculture that relate to solving our impact on the planet are discussed below.

In permaculture we design a property in terms of zones. Zone 0 is the centre of activity; here we focus on energy conservation, temperature mediation, efficient use of space, and meeting the needs of the residents. Zone 1 is the area immediately around the building. We come here once or twice every day for domestic needs and to manage it very intensively. Zone 2 has small domestic stock and orchards. As it is larger, we will have less-intense use of resources, and spend less management time here. Zone 3, further out again, contains our major crops, forage systems, and water storages. Zone 4 is a semi-managed area of forest, natural pastures, and broad-acre pursuits. Zone 5 is the wilderness and natural forest systems.

One aim is to minimise the amount of time and resources we expend moving ourselves and equipment and crops between Zone 0 and elsewhere. This will maximise both our yield and our free time. But the fundamental aim of permaculture is to design Zones 1, 2, 3 and 4 to meet our needs in the smallest space possible, so we can maximise the wilderness. This is because the complex Zone 5 system of forest, savannah, swamp or desert can look after itself better than we could ever hope to, and because it provides incredibly complex buffers for our activities. We need to leave wilderness alone so we can observe it, and learn to approach its subtlety and sophistication in the inner zones.

Few such wilderness areas are left, and these are increasingly under threat - another danger sign at a systematic level. It is very clear that any biological systems' stability depends on the diversity of species, niches and structures in that system. This diversity provides the system with mechanisms for feedback, balance and control which are characterised by varying degrees of coarseness and sensitivity; there are two sets of mechanisms controlling climate, a fast one in the atmosphere, and a much slower one in the ocean depths. We further destroy our land and the huge variety of life forms it contains at our peril.

To successfully manage the negative impacts of industrialisation, desertification, urbanisation and over-population, we need to apply the idea of zones to our towns, our cities, and our trading and financial patterns. We must redesign our systems of production and distribution so that the vast majority of human needs are met locally, from their bio-region.

One of the major problems we face is the excessive flow of resources from many different country areas into the cities. Increasingly our resources are expended in huge transport and trade infrastructures to maintain these cities. We need to set up urban farms as we can no longer afford the middle class demonstration of our social superiority by tending useless lawns and decorative parks and gardens - we need edible landscapes. The reframing of city life so transport needs are minimised and more easily met by mass transit systems is imperative. Trebling parking fees to pay for free public transport would be a start.

We are under increasing pressure to free up our markets to international trade, and to become more internationally competitive. The 'level playing field' argument is based on several assumptions which are systematically incorrect. It is also interesting that most of the advancers of the argument are those who are currently dominant in world trade.

The first assumption is that there is a free market. However, as soon as I have to trade, I am no longer in a free market. If the crop is rotting, or the bank wants its money, I must sell now. At whatever price I can get. A free market is one where there are many buyers and many sellers and all are free to trade. Permaculture systems make sure that there are many subsistence and cash crops. Farm income is not dependent on just one crop, and family survival is not as dependent on farm income.

Another incorrect assumption is that present levels of international transport will continue. International trade will be dramatically reduced when we start to pay the real price of oil in our transportation systems - very few items will be transported more than a couple of hundred kilometres. The systematically ridiculous situation of it being 'economic' to have Jumbo jets flying fresh vegetables around the world each day must stop. The more we limit our purchases of items from distant sources the better, if only to minimise the loss of our economic resource base.

Systematically, the resilience of a system is decreased by taking resources from a weak area to feed up a rich area. Yet we continue to act on the limited economic assumption that we are better off by investing our savings for the highest returns wherever they are available on the world capital market. Practically, you will have more control, more say, less worry (and more fun) from an investment in the local high street than you would have over one in Taiwan. You might even create a local job, and increase the local market for your own products or services.

This leads back to a less formal proposition of permaculture design: start at the back door, now. We can always do some thing today, in our own individual way. Focusing on our local community and neighbourhood can be a good start. Until we change our own habits and perspectives, we cannot expect others to do so.

We must first set up sustainable practices in our homes, before extending our care to larger areas, using only the genuine surpluses we generate from this first step.

These small steps to what will end up as massive social change must become part of our lives soon. The dispassionate technical language of the many reports on these issues does not give this urgency enough clarity or emphasis.

Likewise, how do permaculture ideas relate to the effect of climate change on agricultural practices, those human activities most sensitive to climate changes, and whose effects cascade throughout the social and economic structure of Australia?

The zone analysis and planning outline above is about managing energy within the system. Another aim of permaculture design is to control and utilise the external energies coming into the system.

Permaculture acknowledges our dependence on the natural world, and uses many methods found in nature to moderate the extremes of broad climatic variations.

If climate variations increase due to the stresses we are placing on the system, permaculture-designed agricultural systems will cope more easily with the variation. For example, permaculture dams are often designed to have a 'double depth' section somewhere, to preserve aquatic life, and provide emergency water during drought. Spillways and floodways are wide, planted, and maintained.

Similarly, tree-belts minimise the impact of wind, pests, cold, frost, and fire. Some of the tree-belts are specifically designed with many edible species as emergency fodder for stock and native animals as well as people. Roads, dams and fireproof trees and shrubs are positioned to decrease the probability of fire getting to the house. Cropping methods, dams, swales and banks, and houses and sheds are designed to cope with flash floods and intense storms. Slopes are never denuded, and extra mulches minimise run-off.

We also need to intensify and extend rotational systems for soil building, develop alley cropping methods, establish thousands of kilometres of multi-functional, multi-storey tree and shrub-belts and, importantly, to develop better water control systems. But mainly we need to allow large tracts of the most vulnerable land to return to native systems which we very occasionally forage or graze.

Huge increases in the area of Zone 5 makes sense at an individual level as well as at the system level. The more intensive management required by truly sustainable production systems will mean that we cannot resource the present large areas. More intensive management with fewer imported resources will provide more income, so we will need to crop less area anyway.

As well as ameliorating the effects of climate variations, permaculture designs create more extreme micro-climates. For example, careful placement of shrub and tree-belts can create or displace frost zones on the side of hills. Shelter-belts, slope and sun-traps can allow the growth of tropical fruits in cold climates. Well-designed swales can gather and retain water for tree species having high water requirements. A series of keyline dams can allow sheet irrigation of crops on the slope below, and so it goes on.

Designers use sector, slope, edge, aspect, and elevation to provide widely varying microclimates, which allow growth of substantially different varieties and species from those considered "normal" in an area. The ability to grow different species decreases our dependence on specialised locally adapted varieties. This decreases our economic vulnerability to substantial climatic (or market) changes that make one crop impossible to grow economically, consistently.

The natural selection of appropriate specimens of a species in a number of different micro-climate niches will allow nature to do the analytical work for us.

There are many different techniques, species, and methods of permaculture that are relevant to dealing with climate change. Many more techniques will emerge from careful observation of the results of our design efforts. But the most valuable contribution of permaculture will be based on its starting point, which is to ask "How little do I need?" rather than "How much do I want?" Permaculture goes on to provide many positive ways of meeting those needs that will allow us all to exist without the wholesale collapse of the biological systems we depend on.

Ethics in Permaculture
Should We Use Exotic Species in Our Designs?

Some terminology explained

Endemism, endemic: species that are unique to a localised area.

Translocation: the movement of members of a species from an original specific area which are then released in another area.

Gene pool: the sum of all of the genes of a breeding population.

Feral: organisms which are introduced from other countries. This term often refers to those domestic species which can be found "wild in nature".

Hybrid vigour: a combination of the genes and characteristics of two different species or strains may result in an organism that has special characteristics. This organism may, for example, be better adapted to changes in the environment. A mule, which is a cross between a horse and a donkey, is an example of a hybrid.

Pathogens: disease-causing organisms such as bacteria, fungi, viruses, and including parasites.

Introduction

The encouraging growth of permaculture in local communities has naturally led to its mother science, ecology, achieving greater significance in planning and land management. For permaculture, the broad implications of the application of ecology are complex and difficult to determine, particularly when there are so many different management approaches with an ecological basis which could be applied to permaculture practice.

We suggest that the ethics behind permaculture should be brought to the fore and clarified to prevent a distortion of ecological teachings and as a basis for an appropriate ecological stance. Permaculturists can start this process by identifying the most important current ecological issues and matching them to an appropriate ethic.

As a matter of high priority, the role of endemism in contributing to the ecology of an area and the impact of translocations are fundamental issues. These challenge the integrity of the permaculture movement and need to be addressed. The reason for this is that permaculturists should avoid being accused of contributing to the loss of huge amounts of genetic diversity in those remaining small areas of the planet which are relatively untouched.

Our responsibility

The onus is on people practising permaculture to carefully avoid oversimplifying the solution to our environmental and food production problems. The belief in permaculture that natural systems and people systems have a capacity to recover from environmental degradation through the introductions of foreign species is apt, but not at the expense of losing genetic diversity or having to deal with a host of other concerns by ecologists. An appropriate management response would be to firstly define the remnant values of an area, inclusive of endemic species, and the risk level of dispersal of an introduced species.

Bringing these issues to light is not intended to deny permaculture its capacity for adopting a wider view and therefore greater community acceptance. Ecologists, in general, would appreciate permaculture taking a self-critical approach when assessing the risk of introductions.

From an ethical perspective, we need to remind ourselves of the importance of having an ecocentric view. This means carefully considering the intrinsic worth of all living things and their right to self-perpetuation, free from human interference. Hopefully, being ecocentric in our approach means greater cultural acceptance of permaculture. An ecocentric worldview is something which legal regulation and human-based environmental education programs have failed to achieve.

The impact of exotic species

There are three types of impact: direct, indirect and human-induced. Direct impact occurs through predation of native flora and fauna, the introduction of pathogens and the genetic contamination of local gene pools.

These genetic implications include cases of hybrid vigour and the loss of genetic diversity. One example of hybrid vigour is the introduction of the small fish tilapia. While pure breeding strains of tilapia have been located in two river systems in W.A., the same cannot be said of strains in waterways in Qld and NSW. In these states, tilapia are regarded as noxious pests and translocations of this species is illegal.

In their native Kenya, hybrid species of tilapia have caused the disappearance of several parent species and in some cases even the hybrids, which may be less fit to withstand natural variations in the environment, have disappeared. Our view is that tilapia should not be introduced or translocated, accidentally or deliberately in Australia, but that tilapia culture should be developed in countries of their origin.

A loss of genetic diversity most often occurs with the predation of Australian flora and fauna by introduced species. Some of the well-known examples include the introduced fish gambusia which eats small crustaceans and molluscs, as well as the eggs and fry of native rainbow fish; brown trout and redfin perch, which both have caused declines in numbers of pygmy perch and golden perch; and foxes and cats, which are believed to have caused the extinction of some marsupial species. As a native population decreases in numbers so too does the gene pool of that species.

The introduction of pathogens via translocated species has already had a marked effect on Australian fauna. The major concern at present in the aquaculture industry is the possible introduction of the crayfish plague fungus from either Europe or America. This would devastate the Australian crayfish industry.

Other diseases are known to infect oysters, salmon, silver and golden perch, and yabbies, to name a few. One example is a virus which, being first described in the introduced redfin perch, has now spread to trout and then onto silver perch, Macquarie perch and mountain galaxias. Subsequently, Victoria has banned the translocation of redfin perch. Feral pigs are now implicated in the spread of dieback (Phytophthora spp) which has crippled the Australian timber industry.

Parasites, such as flatworms which attack yabbies or liver flukes in sheep, are increasingly becoming major problems for Government and environmental groups throughout Australia. At present, there are few restrictions of trade between the states, and movement of some species naturally occurs by the interconnecting river systems throughout SA, Victoria and NSW.

Indirect impact occurs when organisms change the habitat or environment around them; there is competition for food, space or light; or food chains and webs are altered.

Habitat modification includes altering water turbidity, soil or vegetation so that endemic species have less chance of survival. The European carp has caused these kinds of problems in the Murray-Darling River systems where populations of silver and golden perch, which generally do not like turbid or muddy water, are threatened.

Feral foxes and cats have changed food chains and webs in much of Australia, where, because of their predation, many Australian mammal populations are endangered or extinct. Other feral animals have altered the vegetation biomass in a given area.

The third type of impact is human induced. This includes a range of activities such as removing feral species which have now become part of a food chain, such as the eagle predating on feral rabbits. The removal of some feral species could be as ecologically disruptive to native flora and fauna as the original introduction.

Other human activities include non-target poisoning, such as the indiscriminate use of herbicides and the destruction of resources due to clearing and development. Often humans cause complex changes to a system, such as deliberate fire, mining and flooding.

From a permaculture perspective, introduced plants should come under greater scrutiny. For years, permaculture practitioners have lauded the advantages of Tagasaste to farmers. It is fire retardant (if cropped regularly), nitrogen-fixing, useful fodder, bee attracting and fast growing. But so too are most Australian acacias (wattles). Maybe we should explore the use of indigenous species more.

Sesbania is reputed to be one of the highest biomass yielding plants on earth. Why is it then that permaculturists go to great lengths to secure seed of *Sesbania sesban* from overseas sources, when they could get local *Sesbania formosa* from northern WA and the Northern Territory? Both species are fast growing, nitrogen-fixing and fodder plants - the difference is that one comes from Egypt, the other is endemic to Australia. Bear in mind that we are not advocating the translocation of this species from northern Australia to the southern states, but again let's see what is locally available first.

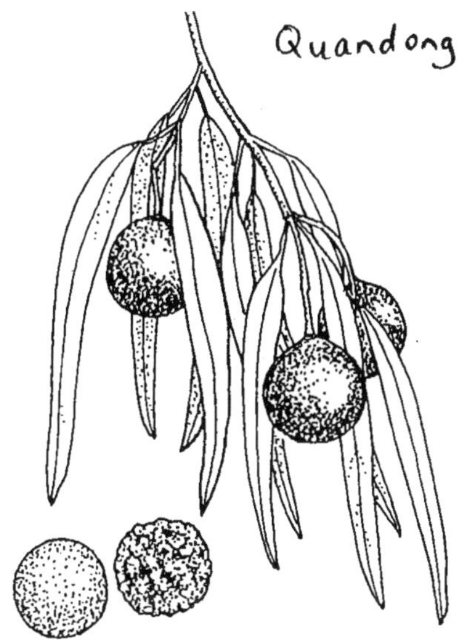

Is it going to be the quandong from WA or the lily pilly from Qld?

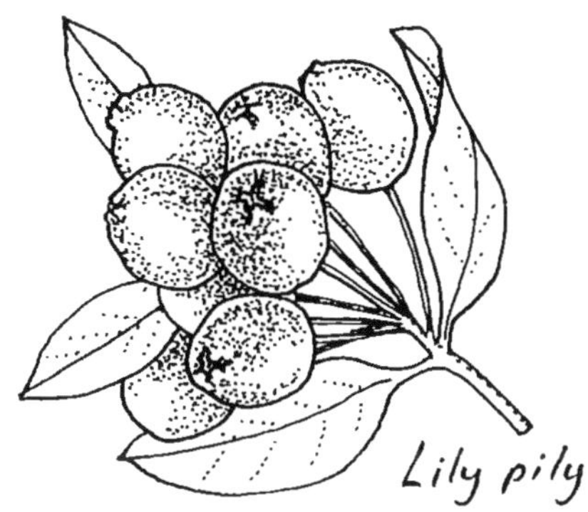

A permaculture solution

The general principles we should adopt, while recognising that some escapes are inevitable, include putting indigenous species first and using them as much as possible in permaculture designs. For example, use pygmy perch in our tyre ponds for mosquito control rather than the introduced, and generally useless, gambusia and goldfish. A second principle is to only transfer fertilised eggs of aquatic animals. This will eliminate the transfer of some of the pathogens and parasites.

Finally, the on-going review of policies for containment, control and eradication, and management is required.

Conclusions

Every foreign animal that has been introduced to Australia has made more negative impacts than positive.

This is true for feral cats, foxes, water buffalo, cane toads, goats, rabbits and donkeys. Rabbits, for example, compete with native herbivores for food, water and shelter; support an increase in the number of feral predators such as the fox; alter plant composition by selective grazing; and disturb soil, which, in turn, leads to erosion and the loss of native plants and weed invasion.

It could be said that the dog and cat have replaced the dingo and possum in what should be a truly Australian culture.

It is not sufficient that the control of the spread of exotic species be the responsibility of the local, state and federal governments. As people who embrace the ethics of permaculture, it is our responsibility to do something positive about it.

Until we have research done on the dynamics of ecosystem disturbance, such as habitat fragmentation, and the effects of changes in fire regimes and rising saline water tables, and until we have an integrated management plan to sustain healthy native ecosystems, we have to seriously consider banning the importation of exotic species and the translocation of endemic species from one part of Australia to another.

Bamboo - A Most Useful Plant

I am constantly amazed how many times we are asked, "What do you use bamboo for?" My first reaction is to say, "How good is your imagination?"

There are the advantages of just growing the plant for soil erosion, windbreaks, habitat and food, but I will briefly explain some of the practical ways to use the timber.

Anyone who has geese will know that ringlock fencing is not an adequate means of controlling them. Add to this a fairly steep slope where they can get a run up and fly over the top, and your problems are increased. Our solution was to weave bamboo culms through the ringlock. Leaving the upper branches on it adds to your barrier above the existing fence, and "bingo" no more flying or roaming geese.

We have also made great use of bamboo for temporary fencing. We frequently establish temporary mulched gardens in our orchard to make use of a space and to control persistent perennial grasses. We often need to erect a temporary fence around bamboo groves so that ducks and geese do not devour the delicious shoots. Bamboo has made this task infinitely easier and extremely cheap. We find the *Phyllostachys* species the best for this as it has two branches at each node. If you leave 2 to 3 cm of these branches on they are perfect for holding the wire down at the bottom and by making a groove on the top of the bamboo stake for the wire to sit in you have a taut fence which poultry cannot get under or over.

One of the most wonderful things about bamboo is it is incredibly flexible as well as being very strong. It's the perfect timber for domes, circles and curves generally - just what permaculturists want. We have constructed the most amazing pigeon dome using *Phyllostachys bambusoides* over what was originally a mandala garden.

This garden already had a 2 m high jarrah and chicken mesh octagonal fence around it so it was a matter of tying two culms together at their tops with baling twine, arching them over a central pole and then securing them either end to the fence. This we did all the way around until we had the framework for the bird netting. We both took a bamboo pole, hooked the netting onto it and then proceeded to pull it across the dome. Hey presto! Secure the mesh by weaving bamboo through it and the chicken wire, add a few bamboo perches, some shelter and nesting boxes and you'll have very happy pigeons that produce squabs almost continually.

Another excellent example of the versatility of bamboo was our recent construction of animal tractors. We always knew that bamboo was the obvious construction material as it is so light and flexible, but we didn't really want to use it on the ground with holes drilled into it to hold the uprights as this would mean water would collect in the internodes and rot the timber. We placed the bamboo into polypipe to get the timber off the ground.

The same principle is used for the dome - the whole frame except for the base is bamboo. We used poly fittings to join the polypipe together and drilled holes in it to attach the bamboo at the base. The nodes proved useful in holding the bamboo into the polypipe holes.

Once the height was established we put the centre and sides on. We put extra bamboo and chicken wire on the lower sides for extra strength and covered the rest with bird netting. Shadecloth and plastic are added on the top for shelter and, in the case of chickens, bamboo perches.

This tractor would be about 3 m long, 2 m wide and 2 m high and is easily lifted by two people or dragged by one. The lightness of this structure was a problem in itself. I'm sure a very persistent fox could lift it enough to get in, so we placed a collar of chicken wire around it which sits on the ground and is secured by, none other than, bamboo pins, again using the branches to hold it secure.

We also use bamboo for marron hides, tepee guards, gates, toys, stakes, shovel, rake and net handles, repairing and securing wire netting and in any situation which calls for the use of timber and less sustainable products such as aluminium, steel and plastic.

If you are using bamboo for timber you should cut the culms that are 3-4 years old and use it green if you wish to bend it into unusual shapes. If you use good quality bamboo, with no cracks, outdoors, we expect it will last from 6 to 15 years. We have a fence around our house made from *Bambusa balcooa* (not the best bamboo timber variety) which is still in good condition after 6 years with only one application of oil when it was built. Like any timber used outdoors, the life of bamboo is prolonged by the application of protective coatings.

It seems like one of the most appropriate plants for Permaculture systems. Quite frankly, I don't know how we lived without it.

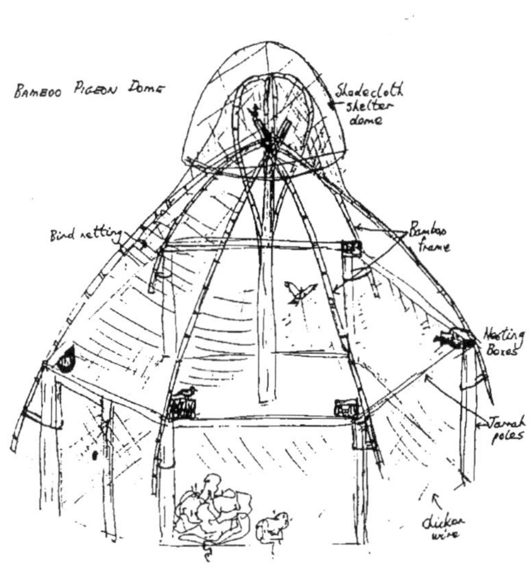

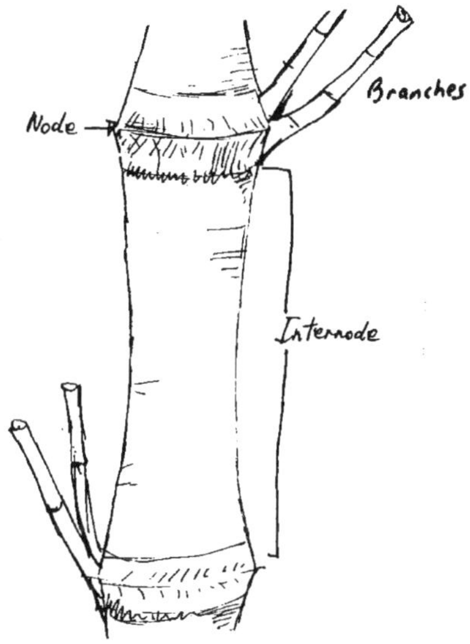

Choosing Fruit Trees

With government planning for our city and suburbs seeming to point to even less square metres per household in the future it is vital to choose fruit trees for the best use of space and productivity over the longest period of time.

It has been my experience that staff at nurseries do not always have knowledge of fruiting times, trial growing conditions and so on, so it is often helpful to tap into the experience of good local gardeners. It's a worthwhile exercise to draw up a chart of times of fruiting of the usual varieties of fruit trees, rhubarb, cape gooseberries, other berry fruits and vines, keeping in mind that if a family can rely on picking even just one variety of fruit at a time, it is likely that their culinary needs for desserts and nutrition needs for vitamins are likely to be met.

I have found that, give or take a few days, it is the same every year for every tree. This knowledge helps in the forward planning for jams and bottling and even holidays, and only by recording and observing can you know about the trees in your particular area. Avoid the feast or famine syndrome!

Even if one has sufficient space to grow the trees, all too often excess fruit can be time consuming, although it is the best of reasons to act co-operatively with neighbours.

Why not use the street lawn area as an orchard - be sure to give the newly planted trees at least one metre diameter of space around each and be prepared to defend it against invading lawn creeping under the mulch. To create interest, maybe each tree can be clearly labelled by a neatly painted sign - many people who buy fruit in a shop have no knowledge of what a plum tree looks like.

Useful fruit trees

Myer Lemon

This is a smaller tree than the usually planted Eureka type, but earlier to mature and produces thinner-skinned fruit which has a golden lemon colour rather than the lime lemon shades of the Eureka. A very attractive shrub for much of the year.

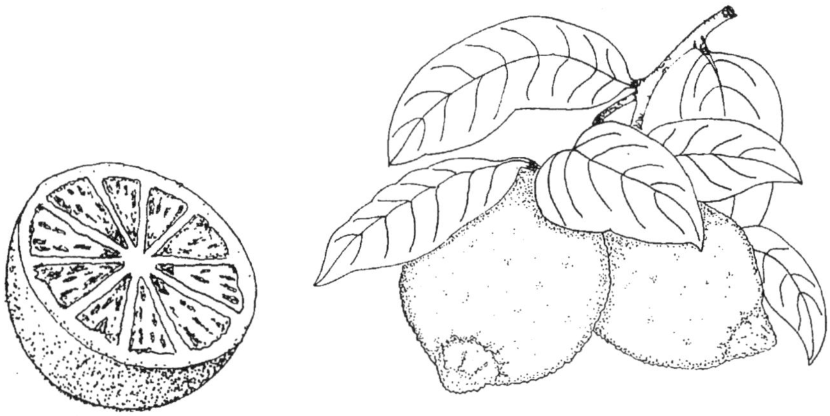

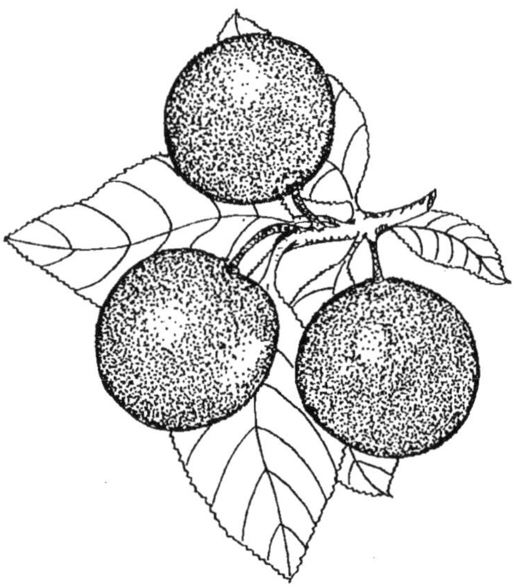

Ruby Blood Plum

To my knowledge this is about the last stone fruit variety to bear in our summer season, thus, birds permitting, fruit can be available well into March.

I have found that this wonderfully flavoured plum, much less tart than Satsuma, can be picked while still very firm, kept in a box indoors for up to three weeks and still the plums are a delight to eat.

Golden Queen Peach

This variety sometimes produces smaller fruit than other peach types but, because of its ability to stay on the tree while the flesh remains firm, it can be used in a more leisurely fashion.

The fruit can be picked unripe and will ripen indoors. The firm flesh can be eaten almost like an apple, unlike many other peaches which, although delicious, must be picked and used within a couple of days, as well as being inconveniently juicy.

The Golden Queen Peach is a heavy bearer and wonderful for bottling, but has the disadvantage of being a "fast growing to enormous" tree and would need regular cutting back where space is at a premium. This tree will fruit towards the end of summer.

Bush Tucker

Ever since Aborigines came to the shores of Australia they have harvested its flora. Aborigines learned which plants were edible and how each fruit, nut, root, stem or leaf had to be prepared and treated for safe consumption. Many plants were also used as bush medicine, and research into the active ingredients of these plants have yielded amazing results. That, however, is another story.

The idea of white Australians eating bush tucker has received momentum ever since a variety of TV specials and several books on the subject were produced years ago. Most people do not consider the native plants that they may grow in their gardens as a source of food. It is fascinating to find out just how many native plants are edible.

Here are some bush tucker plants which are commonly found in nurseries and/or are easy to obtain or collect seed.

Acacia acuminata (Raspberry jam wattle). A small tree up to 6 m. Seeds are collected and ground for flour. The seeds of most wattles are edible. The seeds of the orange wattle (*A. saligna/cyanophylla*) were ground into a flour, mixed with water and cooked as small cake in the coals of a fire. Gum from wattles is also edible and often mixed with water to make a honey-like spread. The bark contains tannic acid and if it is brewed a mild antiseptic can be made.

Acmena smithii (Lilly pilly). Medium-sized bushy tree up to 12 m. Masses of white fluffy flowers in summer followed by edible pink berries. Fruits are eaten raw and they make a tasty jam.

Austromyrtus dulcis (Midyim) . Small shrub that grows to half a metre in diameter. New leaves have a pink colour, flowers are white and the berries are edible. The fruit is about thumbnail size, succulent and is white with purple spots.

Brachychiton diversifolius (Bush peanut). A large tree to 8 metres. Flowers are small and red, and fruit pods develop in spring. The seeds are edible; green when they are unripe and yellow when ripe. Ripe seeds can be roasted and a have a similar taste to peanuts or cashews.

Tiny hairs have to be removed from the seeds before ingestion. The roots of this tree can be baked in hot ashes and have a similar taste to a potato. The white gum can be made into a drink after pounding and then soaking in water.

Brachychiton gregorii (Desert Kurrajong). There are several species of Kurrajong, all of which produce edible seeds. In some species, such as *B. populneum* and *B. gregorii*, the roots can also be eaten. Seeds are usually roasted to burn off irritant hairs and then crushed. The seeds have a high level of carbohydrate (energy) and water content.

*Callistemon sp*p (Bottlebrushes). The red flowers are sucked for nectar.

Dianella spp (Flax lily). Possess evergreen blue-green leaves. Blue flowers and fruit are arranged on a tall (1 m) spike. The fruit can be eaten raw and the leaves can be used to make string and cord.

Grevillea spp - Honey gem. This is a cross between *G. banksii* and *G. pteridifolia* and is well known for the abundant nectar it produces. Aborigines used to suck the flowers of many grevilleas to obtain the sweet sugary nectar.

Haemodorum paniculatum (Bloodroots) Most haemodorums have two to four small flat leaves which are annually renewed, and a red/orange bulb. The black flowers and fruit, which occur in branched spikes about 1 metre tall, rarely open and split to release the seeds. They flower prolifically after a fire. The roots are ground and roasted but can be eaten raw. The taste is mild hot and similar to curry.

Kunzea pomifera (Muntries or Muntari)

Low-lying shrub from the eastern states, useful ground cover with white flowers and edible red fruit (apple-like). It prefers a well-drained soil and is found spreading over sand or rocks. The fruit is found on the underside of the shrub.

Macadamia integrifolia and *M. tetraphylla* (Macadamia). Macadamia is Australia's only commercial native nut tree. The nuts are eaten after washing/leaching and cooking. These seeds have high energy and fat content. Macadamia prefer a warm subtropical climate and a well drained soil. These trees are slow growing.

Podocarpus drouynianas (Wild plum or Emu plum). A large bushy shrub from the south-west of W.A. to two and a half metres, which prefers shade. Large sprays of the attractive foliage are used for cut flower arrangements.

The fruit is purple/dark blue in colour and is up to 5 cm in length. The external seed (green) is attached to one end and is discarded. The fleshy "fruit" (about the size of a grape) is eaten. What is eaten is actually the fleshy stalk of the true fruit. *Podocarpus elata* (Brown pine) is a large tree with similar fruit, from the eastern states. Several species of Podocarpus are found in New Zealand.

Persoonia longifolia (Snottygobble, Long leaf persoonia). A tree up to 5 m high with long thin leaves up to 25 cm. It produces small yellow flowers in summer and the succulent fruit is white or pale yellow, with one large central seed. The fruit dries black with age.

There are many Persoonia species Australia-wide.

Santalum acuminatum (Quandong, Native peach). A small parasitic tree to four metres. The fruit are shiny red when ripe and about the size of a thumbnail. The flesh has a yellow colour. The large pitted nut occupies much of the fruit and has to be cracked to extract the kernel which is eaten raw. Oil from the seeds is used for cosmetic purposes. The kernel is high in energy, protein and fat while the fleshy fruit is high in water and carbohydrates.

Santalum lanceolatum (Bush or wild plum). A small tree up to 6 metres high. The leaves are fleshy, bluish in colour and oval in shape. The yellow-white flowers grow in clusters and the fruit produced vary from red to purple or black when mature. The sweet fruit can be eaten raw. The bark is used as a bush medicine. Both Santalum lanceolatum and S. spicatum are known as Sandalwood. The leaves and wood can be burnt to repel mosquitoes.

Solanum spp (Bush tomatoes, Desert raisins). There are many species of solanums which produce edible fruits. However, there are others which are poisonous as they contain alkaloids. The fruits usually vary in colour from yellow to red and usually have very high levels of vitamin C, and reasonable levels of thiamine (vitamin B1), potassium, iron, carbohydrates and protein. Most Solanums have prickly and hairy leaves and are found mainly in the drier regions of Australia. The black berries of S. nigrum (Black nightshade) can be eaten.

Sollya heterophylla (Australian or W.A. Bluebell). A shrub up to 1.5 m high with stems that twine around themselves and other plants. The leathery leaves are up to 5 cm long. The flowers possess blue petals and the blue edible fruit that is produced can be up to 2.5 cm long with many seeds embedded in the pulp. Flowers October to February.

Thysanotus spp. including T. patersonii (Twining Fringe Lily). Most fringe lilies, so called because of the solitary, fringed purple flowers they produce, have tubers which can be eaten.

The cluster of small tubers from which the annual growth occurs are usually roasted before being eaten.

Our Future With Hemp

Until the 1930s Cannabis sativa, or hemp, cultivation was a burgeoning industry throughout the world. During the 1920's and early 30's Britain, on average, imported 75 000 tonnes of hemp product annually. It was the mainstay of a wide range of commercial manufacturing activities, with Popular Mechanics magazine reporting over 25 000 uses for hemp in 1938.

Prohibition of hemp, due to the expanding use of one plant strain as a recreational drug, has seen this valuable crop dwindle, and in many western nations completely disappear.

However, in an era of rapidly expanding population and unprecedented environmental degradation, hemp is coming back into focus as a major renewable resource.

Many East European and Asian nations, including the rapidly industrialising China, still have an established hemp production base and are reaping substantial profits from expanding global markets. Not surprisingly the West is re-evaluating its worth and trial crops have been established in numerous countries.

Locally, most Australian states are conducting crop trials in conjunction with industry and university groups. Hemp and hemp legislation is under review in these states.

Potential profits from hemp grown in Australia are enormous, likewise the profits from subsequent processing and value-adding achievable by local industries.

From an environmental point-of-view, hemp could be of vast importance in the reversal of farm soil degradation and in native forest conservation.

Growing up to 5 metres in height, it makes an ideal rotation crop, shading out most weeds and dramatically reducing the need for costly herbicide application. This height also provides improved stability in soil evaporation and temperature.

The plant produces between 6 and 16 tonnes of dry stalk per hectare, depending on the strain of hemp and local conditions.

Its superior long fibres are an excellent replacement for cotton which as a crop is widely known for its high dependence on herbicide and pesticide protection, irrigation and for the resultant environmental damage. Nearly 29% of all agricultural chemicals are expended on this one crop.

The fibres are also ideal for the production of high quality paper, though it has been suggested that the coppicing of plantation eucalypts would be less damaging to soils in the long term.

Wastes from fibre production (hurds) also have many uses including incorporation into fibre board and packaging materials.

Hemp-seed oil is another possible venture, with uses in nutrition, cosmetics and industry. Economic viability for fibre or seed oil production would vary but indications are that oil would be the easiest to manufacture, requiring only small farm co-operatives to produce enough raw material.

Fibre, on the other hand, would require a minimum of 1000 hectares under cultivation to support an industry. The higher cost of involved processing capital and lower relative value of the end product is also to be considered.

Australia has the perfect range of growing environments for hemp but with a tardy reputation in terms of new product development and aggressive competition establishing itself world-wide, hemp could easily become another opportunity lost.

Capuli Cherry

The Capuli Cherry (*Prunus salicifolia*) is a hardy, fast-growing evergreen tree that probably originated in Mexico. It travelled to the Andes of S. America where it is wide-spread. It thrives in many different soil types. It grows fast, eventually reaching about 10 m in height. It can begin bearing in its second or third year. It is spectacularly attractive when flowering: great panicles of fragrant white flowers are followed by clusters of long-stemmed cherries.

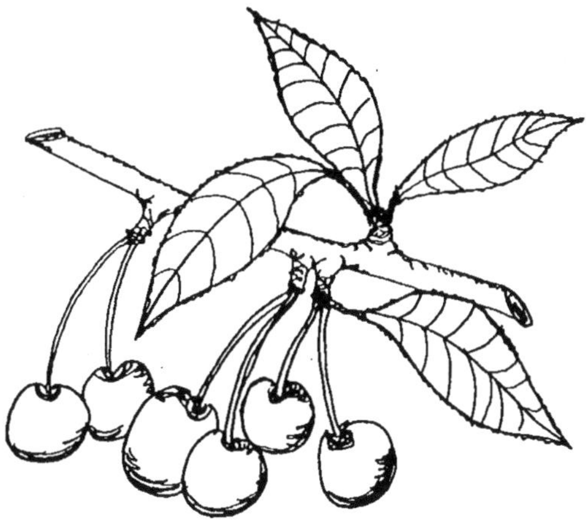

This is a crop that has potential for selective breeding to improve the size and flavour of the fruits which sometimes are slightly bitter.

Timber from young trees is suitable for tool handles, posts, firewood and charcoal. After 6 to 8 years it yields an excellent reddish timber for furniture, musical instruments, and so on. It can be closely planted as a windbreak. It has deep roots which prevent erosion and can be interplanted with field crops. Birds are extremely fond of the fruits, so it can serve as a decoy to (possibly) protect other crops. It is usually propagated from seed and can also be grafted, budded or grown from softwood cuttings.

Moringa oleifera - Drumstick Tree

I never realised the value of the somewhat insignificant looking tree in the corner of my backyard, until someone identified it as the 'Moringa' or drumstick tree. I didn't bother watering my Moringa at all until I discovered what it was. It survived without water but I did observe improvement with the occasional watering. Mine was growing in sand with no addition of nutrients for at least four years that I know of and it still grew rapidly even in my poor soil, needing no addition of fertilisers or pesticide treatment and requiring very little space to thrive.

Apparently it prefers a neutral to slightly acidic soil, but it has recently been introduced with success to Pacific atolls where soil pH is commonly greater than 8.5. It does best where temperature ranges from 26 to 40°C and annual rainfall totals at least 500 mm.

It grows well from sea level to 1000 m in elevation. It will not tolerate long frosts, or extended cold. It will survive short frosts of -1°C or -2°C as an adult tree.

Moringa oleifera is one of the most useful trees for semi-arid and drought-prone areas. Its succulent leaves are harvested for soups, sauces or salads and are quite pale when young, but become richer in colour with maturity. These palatable leaves are high in protein, vitamin A and vitamin C. Where diets lack these essential nutrients - as on Pacific atolls - the Drumstick tree makes a major contribution to human health. Cream-coloured flowers emerge in sweet-smelling panicles during periods of drought - or stress - when the tree loses its leaves. All parts of the tree are eaten widely as vegetables, providing excellent food for both humans and animals (my rabbits adore them).

The pods are often cooked and eaten like green beans, tasting a little like asparagus or chickpeas depending on the variety. The root tastes similar to horse radish (although roots contain high levels of alkaloids) and is a popular food in East Africa. Moringa flowers also produce a good honey. In some parts of the world it is also called the Horseradish tree.

The bark contains a gum that is used as a seasoning - and a treatment for some stomach ailments. Various parts of the Moringa plant are also used in medicines as diuretics and treatments for bladder ailments. The pods are triangular in cross-section - 30 to 50 cm long and legume-like in appearance. The oily seeds are black and winged.

The seed is often used to purify dirty or turbid water. It is pounded into small fragments, wrapped in cloth, and then placed into water jars or containers. This pounded seed acts as a flocculent, taking impurities out of the water as effectively as the alum our water departments use.

Silviculture: Moringa oleifera is easily established by cutting or by seed. Seed can be sown either directly or in containers in spring. No seed treatment is required. The rapidly germinating seedlings can reach 5 metres in one year if sheltered from drying winds and provided with enough water. Plants raised from 1 metre cuttings bear pods from the second year of growth onwards - with maximum production at 4 to 5 years. I have read that in some countries Moringa is treated as an annual and replanted each year, but mine has been growing as a perennial.

Frequent pruning, lopping, coppicing or pollarding will increase and maintain leaf production. Because the leaves are attractive to all livestock, harvesting practices should keep the sprouting area out of the reach of local browsers.

Legume Trees - Permaculture Forage

Trees have been used for forage and fodder by birds, bees and animals since well before the dawn of civilisation. However, unlike shallow-rooted herbaceous plants, they have never been developed as intensely cultivated grazing crops.

Although the family Leguminosae has an estimated 18,000 species, a mere 20 species are used extensively as cultivated crops and another 10 are used as multi-purpose legumes in cool to warm climates. Most of the species used in intensive cropping are shallow-rooted herbaceous clovers. Despite the several real advantages that tree legumes have over pasture species and non-leguminous fodder trees, they are a fairly neglected resource.

Legume trees have multiple uses. Most of them meet their own nitrogen requirements and the deciduous forms return copious amounts of litter, which contributes to mineral re-cycling. Many tree legumes produce herbage suitable for forage by livestock and some produce nutritious pods that can be grazed in situ or milled into concentrates.

Several species can be developed into useful semi-permeable windbreaks and nurse shelter, while others have a distinct role in soil and water conservation. Some produce good hardwood timber useful for fencing poles and posts; others are useful specialist timbers. Practically all tree legumes are good firewood and many are less fire-prone than traditional forestry trees, such as pines. In addition to all their practical uses, many legume trees are attractive and have a high aesthetic value.

Some examples are:

Tree Lucerne	*Chamaecytisus palmensis*
Carob	*Ceratonia siliqua*
Wattles	*Acacia spp*
Tree Medic	*Medicago arborea*
Siberian Pea Tree	*Caragana arborescens*
Yellow Wood	*Cladrastis lutea*
Honey Locust	*Gleditsia triacanthos*
Other Locusts	*Robinia spp*

Acacias in Permaculture

There are between 500 and 1,000 species of Australian acacias. Many species are difficult to tell apart and many are of doubtful validity as species. In addition, many true species contain subspecies or forms which differ in useful characteristics. For these reasons it is better to collect seed from a plant showing the desired habits in an environment similar to that proposed rather than trying to get a plant identified and then ordering seed of that species.

Despite being legumes, acacia seeds are only about 10-15% protein, about the same as good quality wheat. Most acacias contain poisons in the green seeds and pods, but these diminish with ripening and are generally destroyed by cooking. Remember that most food crops also contain natural poisons.

Look for varieties that produce large quantities of large seeds i.e. about the size of a grain of wheat. Large seeds are easier to collect and usually have softer shells and a higher proportion of edible kernel. For human consumption grind the seeds coarsely to break the shell which is then removed and the kernels treated like dried peas or beans. Sheep can pick up large acacia seeds from hard bare ground with their lips while chooks can scratch for and pick up even small seeds from loose soil.

Acacias can be grown in long rotations, being ringbarked after about five years so that their nitrogen is released and they provide a trellis for climbers. They can similarly be used as nurse trees. Acacias can be thought of as substitutes for Tagasaste when the soil is too dry, too hard or too salt, for that species. However, most acacias do not tolerate heavy pruning.

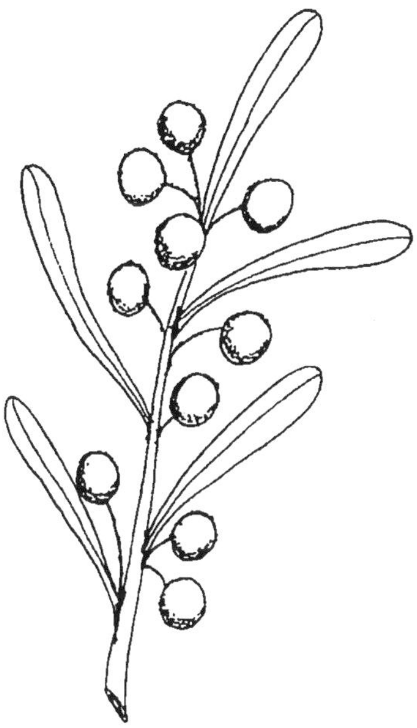

Most Acacias have "hard" seeds with a water resistant coating which must be breached before the seed will take up water and germinate. This coating can be breached by scratching with glass paper, tumbling with sharp sand or scuffing between a shoe sole and concrete. After scratching the seeds place them in a watertight container and pour boiling water over them and let stand. After twelve hours transfer those seeds which have swollen to the germination dish and re-treat the rest. When a small root has developed, plant in 1 litre milk cartons. Plant out with the onset of the winter rains.

Mulch to keep water in and heat out in summer. Acacias respond to fertiliser, but nitrogen is supplied by the symbiotic nitrogen-fixing bacteria which live in the roots. This bacteria is widespread but can be transferred by taking soil from the root zone of an established plant and adding it to the soil of the new plant. Active bacteria is usually indicated by white nodules on the roots.

These types are suggested for food production in winter rainfall areas.

Acacia aneura (Mulga) - small tree from semi-arid regions.

A. acuminatum (Raspberry Jam) - small tree from wheatbelt.

A. longifolia - large shrub. Small to medium seeds, salt tolerant. Good fodder plant.

A. linophylla (Bowgada Bush) - Large seeds, medium shrub. From semi-arid areas. Probably intolerant of clay and grass.

A. retinodes - medium tree (to 10 metres). Large seeds.

A. iteaphylla - spreading large shrub to 4 m. Plentiful large seeds.

A. saligna - tall shrub, large seeds.

The Deadly Difference - What's In a Name?

Some unfortunate plants, like people, have been given bad names due to mistaken identity. Such is the case of a particular plant that often appears in abandoned lands. It is a tough, pioneer type plant, having many uses or functions. A plant that should and will appear in your Permaculture garden (or any other type of garden). Its many functions and uses will be looked at in this article.

This plant's name is Black Nightshade or *Solanum nigrum*, but is often mistakenly called Deadly Nightshade, which is a more poisonous plant. What's in a name you may ask? A lot when your life depends on getting it right. Good science is a bit like a well-written detective story, lots of twists and turns, false trails, red herrings and often an unexpected ending. The deadly difference is also a case of 'vive la difference'. A good free feed is certainly something to write home about.

Placement in permaculture systems

Permaculture systems can be a cultivated ecology. In such an ecosystem little nooks and crannies abound, particularly at the intertwining of the edges of two ecosystems such as the light of a hole in the forest and the shade of the deep forest. This half-sun, half-shade location requires plants with specialised qualities - that is: quickness of growth and withstanding low light levels.

The work of producing such a plant has already occurred in nature. Black Nightshade fits the bill to a tee! It is also self-planting. The agents for its dispersal are birds who eat the fruit and drop the seed contained within on the ground during their constant coming and going from the forest.

It can be assisted in its spread by breaking off fruit laden branches and placing them where they are needed.

Plant description

Black Nightshade is a herb about 70 to 100 cm high. Its leaves occur in threes to nines, ovate to lanceolate in shape. It is dark green in colour with a lighter green underneath. The flowers are white - occasionally with a blue tint. The fruits are green when first formed and turn black when ripe. The flower and fruit are unmistakably of the tomato family, and appear throughout the year. The green fruit should not be eaten.

It is a well-travelled and universal plant occurring in all the temperate and tropical regions of the planet. In a well-established garden it is well hidden, being in the half light of other plants. Abandoned land often has Black Nightshade among its volunteer plant species or weeds; the first flush of growth of the succession of plants which lead to a mature plant cover, be it grasslands or forest. It is not particular as to soil type and does well in most places, to the disappointment of unenlightened gardeners.

Plant uses and functions

Black Nightshade has many uses, both culinary and medicinal. It more than meets the mandatory three functions/uses/products required for it to be placed in a permaculture system.

Culinary

The ripe fruit (black) can be eaten raw or cooked in a pie. Wonderberry Berry is another common name. The young leaves and shoots can be eaten during hard times, when food is scarce.

Medicinal

The plant and fruit can be made into a paste and applied to the forehead to relieve headaches, and it has diuretic and laxative effects.

Other uses can include compost making and covering freshly disturbed soil caused by clearing forest for agriculture. It can be fed to poultry who enjoy the small juicy fruit. It is a most useful and productive plant.

Biological pest control

Not all ladybird beetles are beneficial in the garden. The 28-spotted ladybird beetle feeds on potato and other related plants such as Black Nightshade. In fact it prefers the Black Nightshade to the potato plant. However, it will eat the potato if the Black Nightshade is not present. So if you remove the Black Nightshade, the 28-spotted ladybird beetle will go for the potato plants.

Many pest and disease outbreaks in gardens, orchards and farmlands are caused by thoughtless human activity. The results of such acts shows how little we know about how natural systems work and what they contain. The way to prevent the pain caused by thoughtless acts based on ignorance is by education through protracted and thoughtful, experiential observation.

Black nightshade flan

Permaculture is food for thought, but what about food to eat? Black Nightshade Flan is a way of turning a volunteer pioneer plant into a food plant! This savoury flan consists of a pastry and filling.

Collect the fruit by foraging in your garden, make a pastry casing, and add the berries.

Pastry Ingredients

1/2 cup of self-raising flour

1/2 cup of plain wheat flour

50g butter - diced

water or lemon juice

Method

Combine flours in a mixing bowl. Rub in butter. Add water or lemon juice. Make a stiff dough. Roll out to cover quiche dish.

Filling ingredients

1 medium onion - chopped

1 large potato - thinly sliced

some butter for frying

3 large eggs

splash of milk

pinch of salt

pepper

handful of Black Nightshade fruit

Method

Stirfry chopped onion and sliced potato in butter. Spread over base of prepared pastry Sprinkle on half the black nightshade fruit. Beat eggs and add milk. Season with salt and pepper. Add remaining fruit

Cook in a moderate oven until egg is cooked in the centre. Garnish with extra fruit. Serve hot or cold. Add fetta cheese cut into cubes.

Black Nightshade Flan will be the talking point of your next dinner party. Dare your guests to eat Deadly whoops, I mean Black, Nightshade Flan.

In conclusion

Black Nightshade, while often mistaken for Deadly Nightshade (*Atropa belladonna*), is a useful plant to have in your garden. When it appears in your garden - as it will - give it its due respect and leave it there.

It can cure your headache, increase your urine flow and loosen your bowels. The appetite is improved by eating the fruit. It can increase your sex life, being considered an aphrodisiac. After a hard day being a permaculture recliner in the garden it can put you to a gentle sleep. Provided it's the right plant, it will be sweet dreams!

Celebrity Plant Profiles - Sweet Potato and Cape Gooseberry

Sweet potato (*Ipomoea batatas*)

Keeping the ground covered is an important permaculture idea. We spread various materials over the soil, but there are also plants which will make a living ground cover.

Sweet potato is one plant that will make a luxuriant ground cover and give you an excellent harvest as well. It can survive a long, hot, dry spell although it won't grow. In a moist, delicious garden bed it will flourish like rabbits. Plant them at wide intervals and allow the foliage to rampage over the mulch. Tubers form at the base of the plant and along the vines. The tubers are very nutritious and the tender young leaves are an aromatic addition to salads, soups and stews.

In tropical parts of the world there are hundreds of different kinds of sweet potatoes, different shapes, colours, textures and flavours. In Australia there are usually only two kinds: an orange skinned, orange-fleshed type (high in Vitamin A) and a purple-skinned, white-fleshed kind which has fewer vitamins but a pleasantly different flavour.

Sweet potato is easy to grow. Buy a smooth, good-looking tuber and place it in a container with water half way covering it. Soon shoots and roots appear. When the shoots are about 10 cm long, gently detach them from the tuber and plant them. If there are adult sweet potato vines around you can take cuttings from them. Anywhere along the vine where there are new leaves growing, make a cutting about 10 cm long and put it in a glass of water for a few days. New roots form quickly.

Cape gooseberry (*Physalis peruviana*)

This is a small, shrubby plant (to about 1 m diameter) that grows very well in dry conditions. Plants usually live for several years. They bear fruit prolifically for many months. The fruits are golden yellow, from 1 to 2 cm in diameter, with a tart flavour. They can be eaten fresh or cooked. Dried fruits make interesting "raisins." There is scope for selective breeding to improve the size and flavour of the fruits.

Cape gooseberry is useful as a green mulch supply. Grow plants in and near your gardens and keep them clipped back. Snip up branches with secateurs to make mulch. Use them as nurse plants to shelter some other plant. Keep a cape gooseberry plant in a small pot in your pot plant department and use it as a water indicator: when the gooseberry starts to wilt, it is time to water the pot plants. Chickens adore both the foliage and the fruits, which they gulp down, paper wrapping and all.

Easily Grown Green-leaf Vegetables

New Zealand spinach

New Zealand spinach (*Tetragonia tetragonoides*) is an easily germinated ground-cover plant and is useful on several counts. Firstly, the tips of the branches can be steamed and eaten as a green vegetable, or these tips can be finely shredded as an addition to a sandwich or salad. The plant spreads quickly but in no sense becomes a pest - can be picked right back and fed to chooks and it will grow again - very useful if you have a "no man's land" area of sand. Once established, it needs little water to survive, seems to be pest free and spreads to 2 metres or more across.

Silverbeet

Yes! A common enough vegetable (*Beta vulgaris*) but few seem to grow it. Some new gardeners may not be aware how easily this useful plant is grown - or how long-lasting it can be.

Our first bed of 20 plants were far in excess of our needs but we were able to supply greens for a friend's fowls during a long, cold winter. We found some new ways of cooking silverbeet too, such as a quiche of silverbeet, onion and eggs, and a soup of potato, onion and silverbeet. This plant is very hardy and responds to regular monthly doses of liquid manure or blood and bone. We picked continuously from these 20 plants for 15 months before they went to seed. Silverbeet is generally pest free, except for an odd snail.

NZ spinach

Curly kale

Curly kale

Kale (*Brassica oleracea acephala*) is definitely a member of the brassica family judging by the way the slugs love it! If picked young, the leaves can be steamed as a green vegetable, but by far the nicer way is use the young leaves as for lettuce in a salad - tear into pieces and add diced apple, onion and capsicum before adding French dressing. Our plants gave us food over a period of three months and again the chooks benefited from the older leaves. We saved one plant to let it go to seed - and were amazed at nature's bounty - this one plant first developed hundreds of small yellow flowers which the bees loved. Later these became seedpods about 3 cm long. When the plant dried we crushed the whole plant in a sheet and collected a cupful of tiny seeds - all from one original seed.

Useful Gardening Hints
Jerusalem Artichokes and Perennial Silverbeet

Jerusalem artichokes

It is sometimes said that Jerusalem Artichokes do not produce viable seeds. This is not always true. It is worthwhile to sort through the dried flower heads. There is always a great amount of chaff and undeveloped seeds.

Occasionally there will be a few viable seeds. They look like small, slim, elongated sunflower seeds.

Several years ago I planted a few Jerusalem Artichoke roots bought from a supermarket. They produced about a dozen seeds.

Last year I grew roots descended from the original supermarket roots plus four seedling plants. The seedlings grew very fast and vigorously, and ended up much bigger above the ground than their parents.

When the roots were harvested, one seedling produced a crop very similar to its parents, one produced an even bigger and better crop, and the other two had very inferior crops, merely enlarged lumps at the bottom.

They were all growing in the same small patch, so the differences in crop are probably genetic. Cross-pollination between the seedlings did not increase seed production.

In some areas Jerusalem artichokes can be left in the ground all year, but it is probably safer to dig them in the winter and store in a plastic bag in the bottom of the fridge.

Perennial silverbeet

Silverbeet, usually a biennial plant, can often be forced to become perennial by a lazy gardener. If you pinch out every new flower shoot with great persistence, the plant will eventually stop trying to make flowers and instead will make more leafy crowns on long stalks. The leaves are small but numerous and the flavour is unimpaired.

Successful Bean Growing

Climbing beans - plus climate control

After suffering long, hot summers we decided something must be done about shade for our long, open and west-facing back veranda. During winter we dug the lawn away from the veranda edge to a width of 1 metre, and planted some grape and glory vine cuttings. However, we also decided to try out climbing beans.

We sowed Scarlet Runner, Snake, Freckled Lima and Blue Lake, with great success on several counts. All of these are in need of a high trellis so we used recycled chicken wire fastened to frames or veranda posts and facia boards. Apart from the snake bean I would repeat this exercise each summer, as I believe it possible to grow beans and grape vines almost in conjunction with perhaps just a reduction in the number of seeds planted in order to give the grape vines more room.

The Scarlet Runners are the early ones - can be sown early September and will be bearing beans long before Christmas, forming a dense mass of vines, leaves and beautiful red flowers.

The Snake beans did not bear well here in the south-west and were slow to germinate and grow, possibly needing a warmer climate.

The Blue Lake do not grow such a dense leaf cover, but they bear so well and each bean is truly stringless, which means you don't have to be too particular about picking daily. Also, by liquid manuring fortnightly, and keeping up the water as needed, they will continue to give beans long after bush beans have given up.

The Freckled Lima beans were the most useful in terms of shade. We planted 27 seeds approximately 6 cm apart in mid November. By mid-January they had reached the veranda ceiling and proceeded to thicken out into a beautiful mid-green thatch, and later formed flowers, then small pods containing 3 seeds. These pods take ages to dry and in fact we did not harvest them until mid-May, shelling them for use as culinary beans.

During all this period we were able to have the use of our veranda for a good period of even the hottest day, with coolness, shade, privacy - and food being provided by two handfuls of bean seeds and some old wire.

By the way, the Scarlet Runner beans are perennials and form large tubers under the ground which you can leave from year to year. Ours are sprouting already, in June, so we will have a good start to summer.

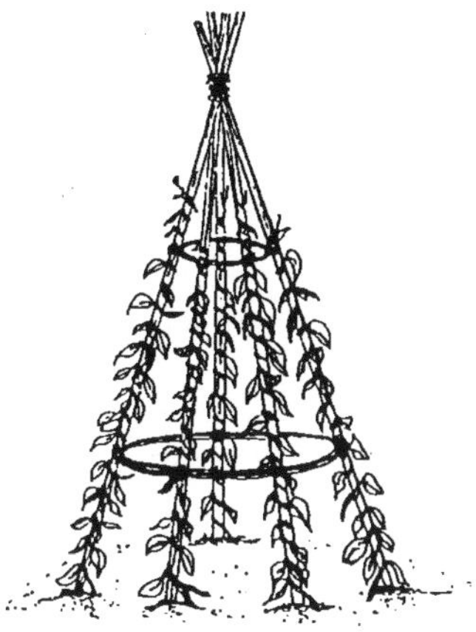

Bush beans

Nothing succeeds like success! You can't fail to have a continuous supply of fresh green beans throughout the whole summer and autumn if you do the following: Sow 15 bush bean seeds in a row about 6 cm apart, water and fertilise as needed. Should be "ready to pick" in 9 weeks. Repeat the above every three weeks from early Oct to early March.

We use "Hawkesbury Wonder" seeds and pick daily. If you find you have too many beans leave them on the bush to dry, and save the seeds for yourself, neighbours and friends and of course the PAWA seed bank.

Green beans picked very young can be eaten raw. Climbing beans are a slightly different scene, taking about 12 weeks to mature but bearing for a longer period.

All these beans can be used for culinary purposes and store well for ages if kept dry in a jar or paper bag.

Abundant Surprises from Permaculture

Imagine an unexpected harvest of field mushrooms in mid-summer right at the back door! This was our New Year surprise. We picked about two dozen saucer-sized mushrooms (11 cm diameter), and ten days later a second flush appeared - this time not counted in dozens but hundreds. Suddenly we were producing mushrooms in commercial quantities. This exercise produced its own dramas - such as eliciting a screech from a bull frog surprised on its evening prowl and unintentionally uncovering a sleeping bobtail lizard under a pile of mulch.

Unfortunately, on the one occasion when some mushrooms were not covered by mulch, a vigorous easterly wind blew up overnight, followed by the first really hot day of the year. This caused some drying out which affected the harvest. Nevertheless, a large quantity of mushrooms of excellent quality were picked the following day. They were generally slightly smaller than the first harvest (at about 9.5 cm) and they were packed up in one-meal lots of about a dozen mushrooms. Such mushroom abundance had me on the phone seeking "orders" from mushroom lovers among neighbours and local group members. All told, we gave away twenty-two lots representing about 400 mushrooms and over ten kilos. Not a bad harvest!

We are still pondering the circumstances which led to this happy state of affairs. The mushrooms appeared on a garden bed along a concrete wall 50 cm above the ground on the west side of the backyard against the dividing fibrocement fence. Passionfruit vines cover much of the fence, while flowering eggfruit plants (about 60 cm high), silver beet, chives and strawberries are happily growing in compost and mulch over a base of sand. This was my first permaculture garden. It has had several additions of organic matter in the intervening period. This bed is behind a patch of native plants of varying heights which block some of the early morning sun and it is shaded by the fence as the sun moves west during the course of the day.

An important trigger for mushroom activity seems to have occurred in mid-September when I prepared garden beds for planting by adding well-matured sheep manure and compost made from lawn clippings supplied by a neighbour and two lawn-mowing contractors.

This mixture was forked through into the soil with a sprinkle of dolomite and the previous mulch replaced.

A few days later, the present vegetables were planted.

We were away on holiday and the garden was watered on four occasions by a PermaLETS member. This less-frequent watering regime meant that the garden was drier than usual.

When we returned home there was a period of warmer days (low to mid 30's) although the nights remained cool and we started watering the vegetables on a daily basis. A few days later temperatures reverted to the generally milder weather we experienced over the Christmas and New Year periods.

Watering was now done mostly every second day but more frequently if needed. I then spread a new load of lawn compost around the plants in this and other garden beds. We noticed the first mushrooms the following day and the second flush appeared twelve days later.

My husband suggests that this pattern of events mimicked autumn conditions - a dry period (summer), followed by warm days and cooler nights with increased rain (initial heavy watering)which stimulated the production of mushroom spores.

In addition, adding compost simulated what is done commercially - a 'casing' of peat moss is spread over the surface once the mushroom body (mycelium) has spread through the growing medium and in 12 days the first mushrooms appear.

Who knows what further surprises may be in store for us.

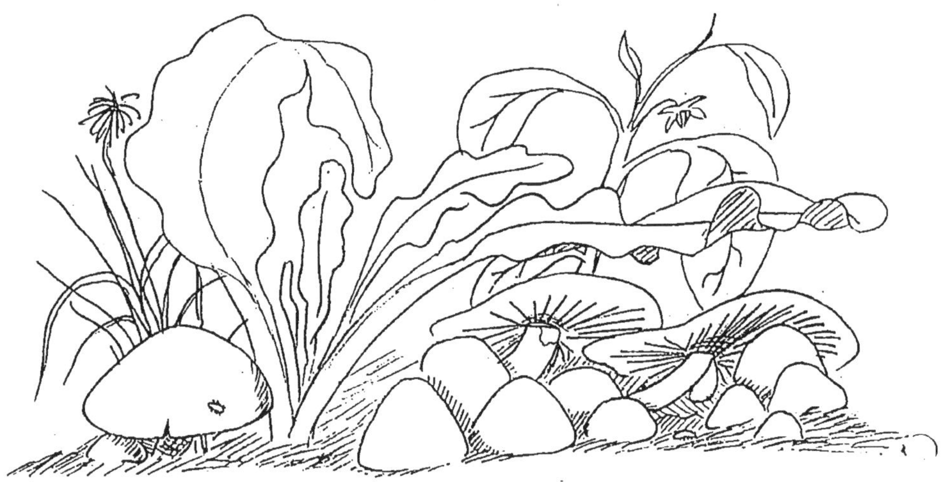

Growing Tips for Vegetables

Scarlet runner bean

How would you like early beans for summer eating and dried beans for winter dishes? Both of these can be produced from the one plant - the Scarlet Runner Bean or Seven Year Bean, as some call it, as it is a perennial bean which forms a large tuber under the soil. Once established it will shoot and grow in its own time each year. However, these tubers can be dug up and eaten in the same way as potatoes, and fresh seed sown for the next crop.

Scarlet Runners are intrepid fast climbers and produce racemes of red delicate flowers which become the beanpod. These should be picked frequently and while no longer than 15 cm. After that they may be too stringy and tough for normal cooking.

We pick the first crop, then keep water and liquid manure up to the plant and are rewarded with a second and even third crop of fruit. If the beans are allowed to grow they can reach 25 cm in length and by letting them fill out and dry on the bush will give you beans to store for winter cooking. Be sure to lay bean seeds out to dry in the sun before storing in a jar.

The pepino

If you are looking for a fruit-bearing small shrub, which is hardy and easy to propagate and produces fruit within twelve months, then the pepino is for you.

This bushy shrub seems to be perennial, easily raised from cuttings if you know someone with this plant - and very often you will find a piece of branch already rooted where it touches the soil. Although it is known as a subtropical it does very well down in areas relatively free of frosts.

Unless controlled it will spread branches in an untidy manner over quite a large area, but by tying it up to a fence and trimming it will become a metre high windbreak (e.g. along the west side of a chook fence).

The pepino is a member of the tomato (Solanum) family and bears purple flowers before producing green oval fruits which become pale apricot yellow when ripe.

Some folk think poorly of the fruit, describing it as "tasteless" and certainly it lacks the tang and aroma of the tamarillo or rock melon but we find the pepino very useful as an ingredient in a fruit salad. Unlike peaches and rockmelons, pepinos can be picked green and take weeks to ripen, and together with two tamarilloes, one orange, one apple and one banana (optional) make a terrific and economical fruit salad for four. Pollination is said to be improved by spraying water from the hose over the bush during flowering.

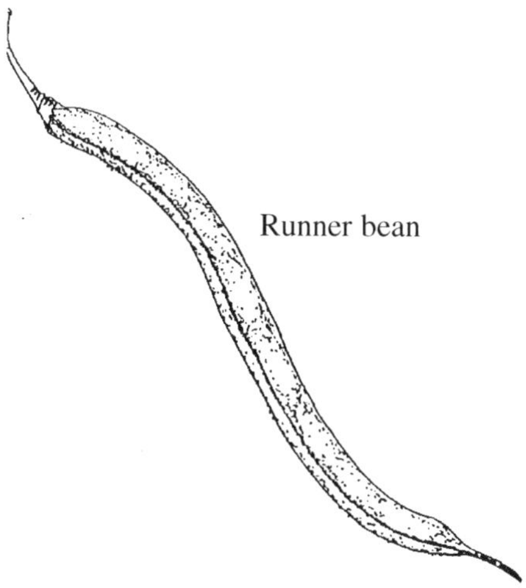

Runner bean

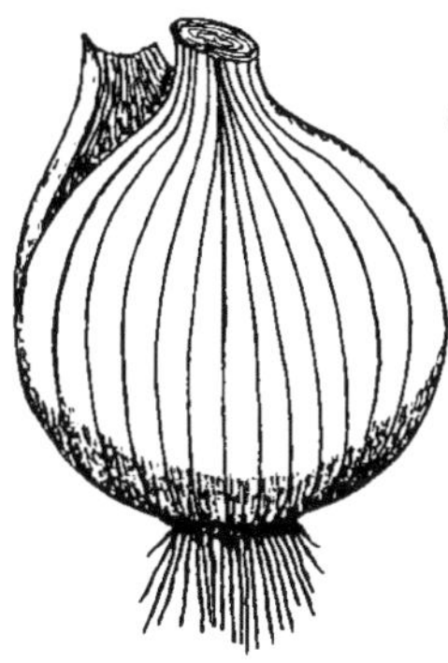

Onion

Allium species seeds

Onions, garlic, leeks and chives will all cross-pollinate, spoiling the resulting seed, as the bees happily go from flower to flower, garden to garden. This may be prevented by placing a small paper bag over seed heads once they flower or removing heads of all but one variety of these plants. Seeds of the onion family do not have long viability.

If you are using commercial seed packets and all else fails - read the instructions. I've had much difficulty in raising leek seeds to seedlings but having studied the writing on the packet more closely, I found that the seed company recommended that seed should be used within 6 months of the sealed alfoil envelope being opened. As few home gardeners need to grow a whole packet of seeds it is a good reason to have a seed sharing buddy and help with costs.

Green beans

Green beans are the easiest vegetables to grow (except maybe radish). A dozen seeds sown at three weekly intervals will give a continuous supply for months. Beginning on October 1st with either dwarf or climbing beans, they can be sown until mid February - even later, if we can guarantee a warm autumn. Last summer we picked our first pods on December 17 (scarlet runners, early cool-weather type) and continued on with dwarf, then climbers, picking our last pods early in June. Any excess was left on the vine to dry before harvesting for storage for next year's seed and for cooking as dried beans.

The productive life of the bean plants can be extended by continuing to feed liquid manure regularly well into the picking time.

Eggplant seedlings

Buy a large, mature fruit. Take a thin slice - look for brown seeds. Bury the slice in a pot, with one centimetre of soil on top. Water it.

Broadbeans

These fix nitrogen on their roots so leave their roots in the ground. Let the last pods remain on the best bushes to dry for next year's seeds. Mulch the bushes as they grow and before their last stages sow carrots between them, beetroots, lettuce or whatever. Let the beanstalks remain to shade the new crop. You might get corn on the sunny side of the plot. Eventually the beanstalks become part of the mulch around the new crop.

A Herb and Salad Picking Bed for Vibrant Good Health

A culinary herb and salad picking bed takes only 2 or 3 m2 of space, and provides unlimited vitamins and minerals for the diet, plus the superb taste treat of fresh herbs in your food, and sensational salads daily.

The bed must be as near to the kitchen as possible. Terraced beds along the edge of the verandah are a good idea as they save wet feet when picking. If the bed is any further away e.g. in the vegetable garden, it just doesn't get used daily especially in wet weather. A good outdoor light needs to be positioned above so that picking can be done as needed, even after dark.

Raised or terraced beds work well. Pick a sunny spot, and water daily in summer. Fill with a rich organic soil and mulch generously.

My culinary herbs include: ordinary parsley, Italian parsley (also used instead of Coriander when that is scarce), coriander, lots of onion chives, garlic chives, lots of shallots poked in any spare spaces, lovage in a shady spot, basil in season and basil mint (in a confining pot) when basil is out of season, thyme and sage in a drier spot, marjoram, oregano (also confined in some way), savoury, lots of lemon thyme, Tarragon (confined by a barrier of some sort), land cress, several English dandelions, several celery herbs, lemon balm, lemon grass (grows large so provides a good windbreak at far end of bed), likewise fennel or dill which need space, and peppermint, spearmint, and apple mint (all confined by pots).

For green salads I grow several oak-leaf lettuces, a red chicory or two, an endive and a mustard lettuce - all of which can be picked several leaves at a time over a long period, along with leaves from dandelions, lemon balm and cress.

Calendulas (pot marigolds) are grown throughout the bed for colour, and the flowers are used for salads. Nearby, nasturtiums and borage also provide flowers and leaves for salads. Rosemary (larger growing and needing drier conditions) should be nearby. Ingredients for the most-used herbal teas can be picked from this 'close by kitchen' bed, but also have fairly close your elderberry (elder flower and lemon thyme tea) and your lemon verbena (verbena leaves and spearmint tea). Vegetables that are picked most frequently should also be positioned as closely as possible. For example, broccoli, when it is grown as a perennial, must be at the nearest end of the vegetable garden, because like strawberries, it needs picking daily.

Also nearby should be the golden zucchinis, the capsicums, and a yellow tomato (which fruits all year round) on a trellis, plus plenty of garlic and perennial leeks and onions.

The above selection, if grown organically and picked daily, will give you really tasty meals and salads, bursting with good vitamins and minerals, and save you money in the supermarket and in doctor's bills.

There are many other herbs and salad plants, particularly for use in stirfry, which could be grown as well. Another health-giving and money-saving addition to your salads is a selection of sprouted seeds such as alfalfa, fenugreek, mung beans and red clover. These are so easy to grow in straight-sided glass jars covered with gauze, so that you have a continuous supply.

Wormwood

Wormwood (Artemisia absinthium) is a hardy, woody shrub that grows to about 1.5 m. The soft, silvery green foliage has a pungent aroma. Wormwood has been used medicinally in the past, and also as a flavouring agent. However, it is now considered to be too addictive and harmful for human use, and should only be grown for other purposes.

Its principal value in permaculture is as a low windbreak, perhaps in conjunction with taller windbreak species. It is one of the few useful plants that are not devoured and destroyed by animals, and helps deter kangaroos from patrolling fence lines. Regular pruning keeps it from becoming leggy and supplies much material for mulch.

Poultry find its shelter quite congenial. It is easy to propagate by cuttings, especially in winter. Just cut a sufficiency of long branches, strip off some of the top leaves, poke a hole in the moist soil and insert the cutting to about 30 cm and firm the soil against it.

If you wish, you can apply a rooting hormone and water the cutting in, but most should "take" anyway. In fact, wormwood is so tough and vigorous that you should consider using only one clone - that is, take all your cuttings only from one plant. This will lessen the chance that it will set viable seeds and start to spread out over a large area.

The Family Umbelliferae

Common herbs dill, fennel and coriander are all from the family Umbelliferae, meaning they have umbel flower arrangements.

Plants from this family are useful in vegetable and herb gardens for attracting bees and predatory insects (insects that feed on, or parasitise, other insects). They are all aromatic herbs and their pungent odours will act as deterrents against unwanted insects.

All these herbs are easy to grow and their soft foliage and umbel flowers make an attractive addition in either a herb, vegetable or flower garden. They easily self-seed.

Dill and coriander are annuals, which means you have to plant seed each year around spring. Fennel is an herbaceous perennial, so can be kept growing longer if you remove the flower heads. Fennel can be divided up to propagate more plants.

Coriander (*Coriandum sativum*)
Uses: Culinary - seeds are a basis of curries and salsas. Fresh leaves give distinctive taste to middle and south-east Asian cooking. Add to salads. Root can be cooked as a vegetable.
Medicinal - aids digestion, improves appetite and reduces flatulence.
Cosmetic - body lotion, toilet waters, aftershave. Add to pot-pourri for spicy touch.
Good companions - dill, chervil and anise, carrots and cabbages.
Bad companions - fennel (coriander and fennel will cross-pollinate and inhibit each other's growth).

Dill (*Anethum graveolens*)
Uses: Culinary - fresh leaves for garnish and sauces in fish and egg dishes, dips, yoghurt, carrots, cucumbers, potato salad, cakes and pastries. Seeds for vinegars.
Medicinal: dill seeds improve digestion and appetite, and sweeten the breath. Dill water relieves flatulence.
Good companions: fennel or coriander, carrots and tomatoes. Attracts bees and repels white cabbage moth.

Fennel (*Foeniculum vulgare*)
Uses: Culinary - fresh or dried leaf for fish dishes, salads, yoghurt and red meats. Seeds for bread-making, fish, sauerkraut and spaghetti.
Medicinal - dried seeds and fresh or dried leaf used for flatulence. Can allay hunger.
Domestic - deters fleas on dogs - crush leaves and rub on dog's coat. Flowering fennel attracts lots of predators.
Cosmetic - facial steam to cleanse pores. Decoction of leaves for eye-strain.
Good companions - dill.
Bad companions - wormwood, coriander, beans, tomatoes, kohlrabi.
All of these plants are invaluable as insect deterrents in the garden, as a flavour additive in the kitchen, and if anyone in your house has a flatulence problem, you should not be without them.

Herb Focus - The Mint Family

The Mint family are well known. Species of *Mentha* have been used for centuries by people throughout the world. Most species are perennial, low-lying shrubs which are shade tolerant (but love the sun), strike easily, love water (and prefer wet, well-drained soil) and spread rapidly. It is because of the invasive nature of mint that it is often kept in tubs or containers. Mints are easily propagated by striking stem segments or by transplanting runners. Here are some common mints and some of their many uses

Applemint (*M. rotundifolia*)
A low creeping variety with white flowers, a slight apple taste and scent, and pale green oval leaves. Useful in drinks, salads, apple sauce and jellies.

Applemint

Spearmint

Typical flower stalk of the mint family.

Spearmint (*M. viridis* and *M. spicata*)

Sweet spearmint flavour, 40 cm, purple flowers in late summer. Green and red-leaved varieties. Grow near roses to deter aphids. Make an infusion for a refreshing tea.

Flavour meat and vegetables, and use to make mint sauce. Tends to die back in winter.

Corsican mint (*M. requienii*)

Tiny peppermint-scented bright green leaves and miniature pink flowers. 25 cm ground cover. Grow around stepping stones in moist places.

Peppermint is good for digestion and for treating colds and influenza. Used to make Creme de Menthe.

Basil mint (*M. spicata var*)

White to pink flowers in summer, 40 cm, oval basil-flavoured leaves which can be added to tomato or meat dishes and in potato salad. Use only a few leaves as too many leaves may cause a bitter taste.

Corsican mint

Basil mint

Eau de Cologne mint

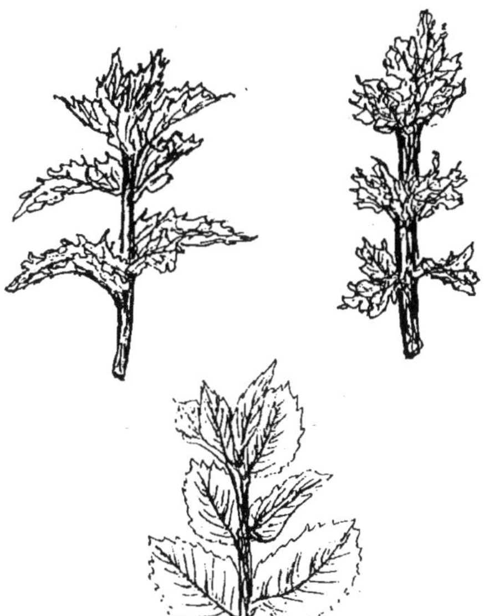

Eau de cologne mint (*M. piperata "citrata"*)

Bergamot-scented (like 4711) purple-tinged dark green leaves, pink flowers in summer, purple stems, 60 cm, loves water.

Good for pot pourri and herb bags.

Chocolate mint (*M. piperata var*)

Closely related to Eau de cologne, spreading ground cover, does have chocolate smell.

Makes a nice refreshing herb tea, and can be used as garnish for cakes and drinks.

Can make a delicious chocolate mint sauce - chocolate, brandy and mint.

Other mint varieties include (L to R) crinkle-leaved black peppermint, pennyroyal and curly mint.

Pineapple Sage

Pineapple sage is a small shrub, similar to most other herbs in the garden, but with one important difference: it spreads aggressively by underground runners. Plant this one near your garden but not in the cultivated beds, or you will harvest nothing but pineapple sage. You can put the pineapple-scented leaves and the brilliant scarlet flowers in salads.

English Dandelion - Weed or Treasure To Behold?

The English Dandelion - *Taraxacum officinale* - is a valuable plant to grow in an area of garden where it can take over without being a nuisance. It self-seeds so readily that it has been labelled a weed.

Fortunately most permaculturalists understand the important role of pioneer plants in protecting the soil from erosion and improving the soil quality to make way for less-hardy plants. There is more than meets the eye with this little plant.

Uses: You can make a soothing cough mixture by covering the flowers with honey and heating to just boiling point. Remove from heat until the honey stops bubbling, repeating this process a couple of times and allow to cool.

The leaves contain vitamins A, B, C, and D, and potassium and iron, so are excellent for using in salads, cooking with other leafy greens or adding to vegetable juices. Tea made using a couple of young leaves has diuretic properties and doesn't leach potassium from the body as conventional diuretics do.

The roots can be lifted in autumn or spring and need to be dug around before lifting as they break easily. Dandelion root cleanses the liver, kidneys and gall bladder.

Two to three year old roots are best for making a healthy coffee substitute by washing them thoroughly and drying them in an oven at 150°C for about an hour (depending on the size of the roots) or until they become brittle and snap easily. Then they can be crushed (roast when needed) and you have a very convincing coffee substitute.

The plant attracts earthworms and is an excellent addition to compost and liquid manures. The flowers attract beneficial insects. The latex is said to remove warts. The essence or healing energy of the flower is for people who suffer from tight muscles due to emotional or physical tension. The flowers are also used to make wine. What more could you ask for from this beautiful little weed? With all these benefits it is just as well that it self-seeds so readily.

Wild Herb Walk

Imagine following a barely existent path that meanders through a variety of sweet scented and useful herbs. Perhaps a frog is hiding under a geranium shrub, or a lizard under a comfrey leaf - it may be real or imaginary.

The initial preparation involves creating a weed-free site. By adding layers of newspaper, manures, weed-free straw and good compost you will build the site up so that drainage is naturally good and the ground fertile.

This is not the time to cut costs or take short cuts as the more resources and effort you put into your garden beds in the beginning, the more you will cut down on costs, maintenance and headaches in the future.

Good garden beds do not require artificial or excessive fertilisers and healthy plants can defend themselves against disease and pests thus reducing unnecessary sprays.

The Wild Herb Walk will always be interesting if you choose self-seeding herbs. But beware of anything that could escape from your garden.

The next step is to decide whether you intend to irrigate or not as many herbs are drought resistant.

Finally, what herbs do you need? There are so many to consider with culinary and medicinal uses.

Some good drought-hardy plants include: rosemary, lavender, aloe vera, geraniums, marjoram, oregano, thymes, wormwood, hyssop, St John's wort, evening primrose, wood betony and southernwood.

You should harvest drought-hardy plants before the seed drops, so that the plants stay contained in your garden. Fennel is a good example of how herbs can get out of control if it is allowed to self-seed.

Irrigating increases the range of herbs that can be grown. Unless you live near a stream, the plants that require moisture will not escape from your garden, therefore you can allow them to self-seed. Prolific ground covers such as lebanese cress should not be grown near a stream.

Ideally the meandering path of the Wild Herb Walk should lead to a comfortable seat surrounded by scented herbs. Whilst sitting you could ponder what herbs to harvest on the trip back. Perhaps a herbal tea is awaiting you, or a deluxe meal seasoned with your own herbs. Or maybe you'll just sit there and dream away.

Soil Fertility

Soil mixes that contain a balanced ratio of key ingredients can markedly reduce the need for large volumes of compost and water. The need to add large and continuous quantities of compost to many soils to sustain fertility is very labour intensive, requires ongoing soil disturbance and can present unwelcome opportunities for cockroach, ant and rodent feeding.

The extra effort and cost of finding out (or testing for) what is missing from your area's soil is well worth it. Because soil amendments such as rock dust need only sprinkling on the soil surface and watering in, there is little or no need to disturb the new soil mix.

By starting off with a soil mix that approximates a forest soil, the further addition of compost or amendments can be markedly reduced because they will be better retained by the high carbon 'sink' (woody/lignin matter content) and not be leached out. Because most people overlook the need for high quantities of woody matter in the soil they become tied into an on-going labour-intensive cycle of large-scale nutrient addition.

Placing this woody material on the surface only, does not approximate forest soil or stop the downward leaching of nutrients. It also introduces other problems such as increased water requirements, due to higher evaporation from the mulch.

Below is a basic soil mix recipe.

- 30% (min) hardwood branches and/or logs (or very coarse wood chip)
- 15% manure
- 30% sand (or a mixture of sand and loam)
- 15% (max) clay
- 10% leaves and twigs

The exact ratio of ingredients for a soil mix is relatively critical in relation to wood, sand and clay content but can be varied quite a bit in relation to manure, loam, leaf and twig content. These more critical ingredients are varied only slightly for plant type, depth of garden bed or tree hole and, more importantly, the surrounding soil type. Sandy soils will require the highest (15% max) proportion of clay and about 40% woody matter and 20% manure.

A heavy clay soil will require no clay in the mix but an increase of the sand to 45%. The virtues of these soil mix ratios are best understood by describing the special attributes each of the materials have.

Woody material

Woody material is a must in most soils and is often totally overlooked by the excessive attention given to mulch and compost. Within a year, mulch and compost can totally disappear into the soil, as it breaks down into very fine compounds. This often does only a mediocre job of improving the soil structure and long-term soil fertility. Large annual applications of compost and mulch are then required to keep soil health happening.

A quick observation of nature reveals that the backbone in many soil building processes are the activities of perennial acacias and other short-lived woody herbs and shrubs. When they die, they leave vast quantities of woody matter in the form of dead roots behind, but even as they live, dead sticks, seasonal root death and a huge variety of very fibrous woody pod material is constantly returning to the soil.

In sandy southern regions the banksia is also a major contributor. This woody material, unlike grass, compost or mulch, is particularly rich in lignin. This resists breakdown and is very attractive to wood mould, soil fungi and other micro-organisms which, like many of us, prefer a more permanent home than an annual tent city.

Once established within such a semi-permanent home, these soil organisms get busy creating many compounds that our composts and mulches would be hard put to match. The woody matter retains moisture better, distributes it better and can easily persist for several years.

Also let us remember that the fungi are often the micro-nutrient generators of elements such as zinc and phosphorous and directly couple up to the root hairs of plants to act as nutrient corridors. The deep, black, open, soft soils of old forests are more a product of lignin (woody matter) and its extensive organism colonies than compost.

By growing extensive root systems under the ground that annually and eventually die back, all plants (but particularly acacias and woody herbs) markedly increase the long-term below-ground levels of carbon. Many nutrients are chemically snared by this carbon 'sink' that stops them being carried deeper by leaching and allows them to be released slowly over a long period with the assistance of soil organisms.

The comparison between refined starches and sugars, and fibre-rich foods is as appropriate to the soil as it is to ourselves. At least 30% woody matter in the soil will save a lot of extra feeding of nutrients and water to our gardens.

Clay

Clay readily absorbs water but not air and so can be said to be very poor at breathing. It can be very high in iron or aluminium and often harbours numerous other nutrients. When dry, it can set like concrete. The setting action of clay is an important consideration because it is mobilised by water and spreads easily through a soil mix, rising to the surface if ponding occurs. Upon drying, this clay skin can impede air transmission and interfere with water percolation down into the soil, causing it to be lost to surface evaporation. Like woody (carbon) matter, clay holds water and acts as a sponge for many nutrients that are lost from loose soils on mid and upper slopes. Its very high mineral content can either provide nutrients or lock up other necessary but less abundant ones. Obviously, when all these items are considered it is clear that clay can easily go from being a friend to a foe. It is because of these characteristics that claypan environments are only able to host specially adapted plants on their surfaces, but within a short distance of their margins a myriad of plant types flourish, making use of its benefits but being kept back by its extreme self-preserving qualities.

Keeping the clay content below 15% of the total soil mix avoids the concreting action of clay while maintaining its benefits as a water and nutrient reservoir. Clay also acts to give the soil more 'ionic' character, thereby helping water to mobilise nutrients.

Sand

Sand is made up of very small, angular grains that don't cohere to themselves, but rather stand apart from their neighbouring grains, creating many small cavities throughout its volume. Because of this, sand readily holds air but not water, which easily percolates downward through it. So one could say that it is complimentary to clay in that sand breathes air and clay breathes water.

The well-draining capacity of sand is crucial to plants, many of which need to have access to small regulated amounts of air at their roots (as do the myriad of soil micro-organisms) or they will suffocate and die. Many gardeners do not appreciate the incredible benefits to many plants of a 'wet - dry, wet - dry cycle', in which they are alternatively supplied with water and then air.

Even bog and emergent water plants can be kept healthy on this sort of regime with vast savings in water. The air invasion assists in the oxidising liberation of nutrients and rapid expansion of organisms that boost fertility even higher.

The flooding action of water then kills many organisms and mobilises these oxidised nutrients into more complex compounds and on into plant roots.

Sand provides the right level of drainage that allows such a rapid turn-around between air and water. Obviously with too much sand there can be little or no water retention and severe nutrient imbalances, so it is only used in larger quantities if the surrounding soil is heavy with clay, to open it up and improve its drainage.

The lightweight and loose nature of sand allows even small plant roots to push through it so it is a very useful addition to heavyweight rocky soils. These loose and non-cohering characteristics allow it to flow into large air cavities between the other materials, reducing these to very small pore spaces that roots can operate in without drying out severely. Sand also introduces good sources of silica, iron, calcium and often manganese, boron, titanium, vanadium, copper and zinc among other elements. It also provides the 'grit' that chooks and earthworms utilise in their gizzards for grinding action.

Manure

I would prefer to use the word fertiliser but this has become so synonymous with chemicals that I prefer the language of gardeners. Manure is therefore a very broad term which is applied to the waste of animals and, in general, to any dead organic residues we feed to plants. In the context of this article we use the term to include dead plant material like waste seeds, pods, compost, dead insects and actual dead animal bits as well as their excrement. In short, just about any natural concentrated animal, insect and plant compounds.

Manure contains many elements and complex compounds in an organic form that break down into readily-mobilised nutrients that are easily absorbed by plants and especially soil micro-organisms. Manure is often very rich in micro-organisms itself, acting as a major soil fertility source.

Being high in nitrogen, phosphorus and sulfur (which are often low in soils), there are generally immediate responses to such an addition by plants hungry for such elements.

The nitrogen also helps offset the high carbon levels of the woody matter, which becomes particularly dynamic at around ratios of 30 carbon to 1 nitrogen.

Around this ratio (30:1) colours darken, microbial activity explodes and decomposition of the carbon accelerates, indicating how a little nitrogen can go a long way. A good example of this is if you add urine (nitrogen) into a bucket of sawdust (carbon). The colour of the sawdust will steadily darken and its capacity to behave as a 'sink' is quite good until you get to this magic ratio and then once past this point it can no longer absorb the nitrogen.

By then it is quite dynamic, bordering on toxic, as the beginning of ammonia odours will testify, at which point it is fed to the garden. Ash and charcoal are even richer sources of carbon and so are used in pit toilets as a 'sink' to slow microbial activity and thereby suppress the resultant odours.

This raises another issue; that too much of a particular compound can suddenly result in toxicity. So once again it is important to keep an eye on the mixture's ratios. Because there are relatively high levels of woody matter used in the mix there is a much higher tolerance of excess manure or rotten fermenting substances.

Because of the wide range of items we refer to as manure there is a huge range of differences in what they contain and how slow or fast their breakdown is. Dead insects, fish, other animal bits and offal are excellent manures but can be problematic with dogs and foxes as they may dig up a new garden looking for bones and rotting meat. Burying them at the very bottom of a tree hole, layered vegetable garden or at least lower in a soil mix should alleviate this problem.

Last, but not least, is the assertion that too much woody matter will starve the plants of nitrogen. Whilst this argument is complex, I am yet to be convinced that nitrogen is the cornerstone of anything other than the chemical industry's way of forcing plant growth at the expense of environmental health. Plants subject to such nitrogen poisoning are forced to take on lots of water and suffer chemical imbalances that weaken them. Their fruit and produce can't let go of this moisture and so have a very short shelf-life that ends in them rotting rather than shrivelling. Their flavour is 'watery' and weak and their nutrient value is poor. The effects of nitrogen poisoning of the soils, plants, animals, food supply and streams are all around us, for those with eyes to see.

A certain amount of nitrogen is needed to give the micro-organisms a boost and to offset any immediate impact of excess carbon. Within the first few weeks, these soil organisms set up a more balanced soil chemistry that the plants can contribute to and, from there, all the soil chemistry and soil 'stakeholders' use the carbon to manage their own environment. In fact, the carbon acts as a buffer to any toxicity and wayward chemical reactions in a way that low carbon soils never can.

Leaves and twigs

These materials provide many complex chemicals, microbial colonies and soil organism foods. They get the decomposition process under way quickly and with a good variety of organisms. Leaves and twigs are halfway between manure and woody matter in their useful lifespan, being relatively exhausted within a year. Their other effect is to spread moisture sideways and more evenly through loose or sandy soils thereby slowing downward percolation losses.

All of the key materials together synergistically enhance the performance of all the others and this lays the foundation for long-term fertility with minimum effort.

Volumes required

The main question now is how much of this soil mix is needed? One could reply that all your soil needs to be like this. However, this is not realistic or generally necessary. If the plants can survive to their third or fourth year without having to substantially deal with the surrounding soil then they are by this time relatively robust and healthy enough to deal with it. By this time also, the organisms have reached out and started to alter the immediately surrounding soil. For most native plants and hardy exotics, just supplementing 40% of the planting hole's volume with woody material, some manure and 60% of the original soil will be more than enough to get them raging. The square hole size for most of these type of plants need only be 0.3 m square x 0.3 m deep (0.027 cubic meters).

For more demanding food-producing exotics that are not at home in your area's soil type, a much larger volume of basic soil mix will be required for each of the two extreme soil types (heavy clay or sand). In heavy clay soils we have found that a minimum square hole size of 0.5 m square x 0.6 m deep (0.15 cubic meters) is good enough for deep rooting trees. For palms or surface-rooting plants the square hole needs 0.6m sides x 0.4 m deep (0.144 cubic meters).

When surface-rooting plants have restrictions to the sideways spread of their roots they very often become restricted in their upward growth, so wider rather than deeper is required. It is important to get at least 0.1 cubic meter of this soil mix to give the exotic food plants and trees a healthy start in life.

In sandy soils that are also substantially easier to dig, it is ideal to create a larger volume of soil mix to grow into, as fertility and organic levels are generally extremely low. This is further exacerbated by the free-draining nature of sand that causes downward leaching of any soil improvements. By adding 100 mm or 0.1m to all the above hole dimensions, you get a dramatic increase of over 1.5 times the volume of soil mix, which would be adequate for sandy soils. Remember that this is a minimum volume aimed at giving the young plants a 3 - 4 year head start in life so they are in a more robust condition to deal with problematic soils.

Existing plantings and broadacre

Obviously these single-plant techniques are not necessarily ideal for someone who has already planted out their plants before reading this article or someone working on a larger scale such as an orchardist or broadacre farmer.

For these people there are a number of techniques that may be useful for raising the woody matter content of their soil or incorporating this soil mix where plants have already been planted.

Getting a soil mix into the ground adjacent to recently planted trees may require narrow (0.2-0.3 m) trenches alongside the young (1 - 4 year old) trees within 0.3-0.5 m of their trunks, backfilled with the soil mix.

The aim is to get as close as possible without damaging their main roots so extra distance and care needs to be exercised near surface rooting plants such as citrus, palms, figs, and some berry and pip fruits. Some exploratory manual trenching is advised to avoid breaking main roots. Trenching on a parallel, square or triangle pattern around each tree will help its developing roots get into the mix as early as possible. By auguring 3 or 4, 0.3 m holes around each young tree you will achieve the same results. Once again the depth of trenches or holes for surface and deep rooting plants is 0.4 m and 0.6 m respectively, (unless in very sandy soils and then it would be 0.5 m and 0.7 m respectively).

Ultimately, the addition of woody matter is the primary objective to establish long-term fertility. In true permaculture fashion we will do it best (and easiest) when we observe and copy nature's time-honoured technique of 'growing it into the soil'.

The selection of woody herbs and/or acacias that are quick growing and short lived (annual, biennial or sacrificial crops), to grow between existing trees will achieve such an aim. However, this does require some planning and information on which plants will best perform this task in your region (such is the nature of permaculture).

Things to look for besides speedy growth and short life are the hardness of their woody material, their root system profile, height, width and possibly their flammability. It would not necessarily be an advantage to select tap rooting, sparse rooted or surface rooting plants but rather plants that have extensive midsoil root systems.

Some deep ripping prior to seeding could assist them to really spread out into the soil. If the plants die naturally after a short time this is most probably more ideal than having to go to the trouble involved in having to kill them off. Depending on what vehicle, animal or other access is required for the area under consideration, the layout and duration or cycle of this soil improvement strategy can determine the ideal plant form and lifecycle. Once the ideal plant form and root shape has been selected along with the desirable lifecycle, there can be further selection of specific plants for future products, fodder or functions, such as fire resistance.

A good example of plant families that meet many of these criteria are the Chenopodiaceae (Atriplex, Rhagodia sp.), Goodeniaceae, Malvaceae, Zygophyllaceae and Surianaceae. From these plant families can be selected a wide range of woody herbs, all with a very wide range of individual characteristics suited to your region's climate and soil types. Many native plants in your area are busy functioning in this manner already and they may be suitable.

Further soil improvement can be gained by growing plants that are dynamic accumulators of the deficient elements and feeding them the required amendments when they are in their seasonal growth flush then turning them into the soil as a manure crop.

This way they carry the amendments (such as rock dust, compost or manure) into the soil as a converted set of complex compounds already more enhanced than the initial application. This turning-in activity could also coincide with the sacrificing of any plants grown for their woody matter.

Two and three year fertility cycles are best contemplated if there is a large area to be managed in this way, so the biennial woody herbs and accumulator plants become a very useful group to select from if machinery is unavailable or undesirable. Also the introduction of some type of grazing animals to feed on any multi-use plants will have the added benefit of manuring and trampling if the fodder levels and stocking density are high enough. The level of complexity is limited only by your knowledge of plant and animal characteristics and their suitability to your region.

A cautionary note: Don't design and implement things you've only read about without running some trials and scouring the region for local information about the plants and animals you are contemplating.

The eagerness to do everything at once without extensive background knowledge is not good practice. If local knowledge is lacking and you're sure you've got a good idea, set up two or three different small trial patches first to establish the management implications.

Each trial patch needs only to contain some very simple combinations of 2 to 5 plants that you trial for the first year or two to see what the interplay is between them and the effects on the soil.

Remember, if you get it right, the information you gain will accelerate your soil building, making those first couple of years well worth it. However, if you get it wrong on a large scale you could easily spend the next decade trying to control some feral plant you've introduced or dealing with irate neighbours who can destroy your business and lifestyle because you poisoned their stock with an introduced toxic plant. Once again, there's that critical permaculture principal of observe, observe and observe.

Soil Life - Buried Treasure

Study the sizzling topic of soil deeply enough and you will soon find yourself delving into the world of alchemy, astrology and magic! Soil responds to cosmic and celestial forces, atmospheric electrical currents, paramagnetism and energy (including thought and sound) radiations.

Under appropriate conditions, soil can demonstrate an increase in the levels of particular elements that is totally inexplicable to modern science. Indeed, modern science has hardly begun to look beyond soil as a simple substrate in which plants anchor themselves.

Soil life, for example, controls everything to do with soil fertility and productivity except rainfall and temperature and yet we know precious little about it. Indeed, understanding and managing soil biology has the potential to reduce the use of expensive and environmentally-damaging inputs to agriculture and improve food quality.

What are soil biota anyway? Most obvious to the naked eye are the macrofauna - big critters such as earthworms, millipedes, crickets and earth-dwelling stages of various insects (eg beetles), and the smaller microfauna such as mites, springtails and termites. Then there are various plants, fungi, protozoa and bacteria, most of which have not yet been identified and classified.

All soil life has a part to play in recycling organic residues such as manures and plant matter back into the soil. The contribution of earthworms in transforming organic material into higher class, accessible nutrient is well recognised.

Similarly, microbes involved in composting combine carbon-rich plant residues with nitrogen-rich manures into a plant nutrient of higher quality than either of its 'parents'.

Soil biota also transforms non-organic ions into forms more available to plants, accounting for the occurrence in soil of essential nutrient forms such as sulfates, nitrates and phosphates.

The stable soil organic matter created by soil biota is an important determinant of soil texture, aeration, nutrient-holding capacity, water infiltration and holding capacity and pH buffering ability.

The mycorrhizae group of fungi form a symbiotic relationship with plant roots and can offer them much extended access to many minerals. One experiment, for example, showed mycorrhizal fungi afforded a nine-fold increase in plant access to phosphorus!

A famous group of bacteria are the rhizobium, capable of fixing atmospheric nitrogen in legume root nodules. Inputs for some clover associations would be in the order of 200 kg of nitrogen per hectare per year - by far the cheapest way of replacing that lost to production. Other bacteria facilitate mineral weathering processes, releasing plant available minerals from the existing soil and from applied natural mineral fertilisers.

Soil microbes need comfort to function well. Comfort comprises the right moisture, temperature, air supply and nutrient supply and an environment free from chemical insults.

Acidic soil conditions, for example, harm earthworms and the rhizobial bacteria. Artificial fertilisers can suppress microbial populations and the enzymes that they produce (these are key factors in nutrient recycling), rendering the system increasingly dependent on chemical inputs.

Enhancing soil biota performance through comfort can be achieved by the following means:

- Maintain adequate mulch cover on the soil to prevent drying and temperature extremes.
- Investigate and rectify base mineral imbalances. e.g. through appropriate prescriptive liming.
- Use green manure crops to increase soil organic matter (to be effectively utilised by soil critters, surface organic material must be turned into the soil or at least cut, chopped and covered with a light layer of soil using a rake or scarifier).
- Avoid burning.
- Aerate the soil. A light, non-inverting forking of the garden, or use of a mechanical aerator on acreage will help air to penetrate soil and also improve infiltration of water and retention of silt, manure and seeds on the soil surface.

Look after your soil critters and the rest will take care of itself. Happy growing!

Farming For Soil Health

We often tend to overlook the connection between soil health, the health of crops and livestock and the health of humans consuming them. Human health is a vital component of an organic approach to farming, which is ultimately linked to soil health. The current continual use and application of superphosphate on our farms is having serious effects on the soil.

Soil management practices can increase the amount of available phosphorus significantly without the application of any extra phosphorus to the soil. Much of the phosphorus in super, however, becomes unavailable to plants when it combines with calcium, iron and aluminium to form the highly insoluble mineral apatite.

The phosphorus in apatite can be released in small amounts by the action of bacteria which are dependent on sulphur. Improved soil structure also contributes to greater phosphorus availability by holding the phosphorus in the soil, bound to clay particles and humic acids (derived from organic matter), which will readily release the phosphorus to plant roots on demand.

Superphosphate is very acidic, so making the soil alkaline also increases phosphorus availability to plants. Alternatively, the addition of gypsum (calcium phosphate) to heavy soils has many advantages. Gypsum is relatively neutral and is an excellent source of sulphur in acidic soils, as well as changing soil structure to allow better water penetration.

Two of the elements most readily rendered unavailable by superphosphate are copper and zinc (these form insoluble phosphate compounds). This leads to serious deficiencies in soil, particularly in those ancient and nutritionally-poor soils found throughout much of Australia.

Both copper and zinc play a vital role in animal health. Copper is essential to the production of haemoglobin and keratin, the protein of fingernails, hair and wool. Zinc is vital to the health and strength of skin. As with humans, animal health is related to a range of factors, nutrition being one of the most important ones.

And just as humans and crops are more vigorous with good nutrition and more resistant to a host of diseases, so too are livestock less disease-prone, and this requires healthy soil with a well balanced supply of nutrients.

Organic farms often lack an over-abundance of weeds. Many weeds are pioneer plants, adapted to low soil fertility and an early stage of succession.

These plants have an important role to play in establishing cover over bare soil, reducing erosion, contributing organic matter to the soil and recycling nutrients leached into deeper horizons of the soil. Weeds can act as useful indicator plants, providing the observant farmer with information regarding the state of the soil. In healthy soils, many weeds lose their 'competitive advantage' over crops, although any farming system based on annual crops invites the growth of weeds for that period when the soil is relatively bare.

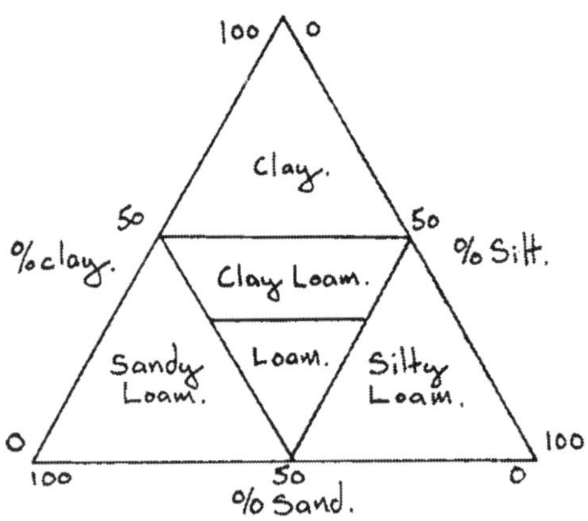

Not only are organic methods proving productive, they are also proving to be attractive in the market place. Organic wool is in high demand in Europe due to the stringent effluent quality requirements for the wool processing industry. Organic wheat attracts a good price and contains a higher protein content than most chemically-grown crops. Organic wheat also has a higher ratio of grain to husk, meaning that the same weight of grain will produce a greater weight of flour.

The narrow conception of soil fertility which has dominated agriculture to date, and a lack of understanding of soil drainage patterns, has led to a severe decline in the fertility of our agricultural soils. Re-establishing soil health and fertility will re-establish the health of plants, animals and people.

The Big Nitrogen Fix

In permaculture design, the use of plants that, in association with bacteria, make atmospheric nitrogen (as nitrate) available is common. It is often assumed that there will be a supply of nitrogen available to other plants that grow nearby. Does this actually happen? If so, considerable saving would occur as the cost of artificial nitrogen is very high, a great deal of energy is needed and pollution is often produced when artificial nitrogenous fertilisers are made. What does this mean for permaculture people?

Some plants can do it

Not all plants have this marvellous ability to fix nitrogen. Legumes are a well-known family of plants that, in association with bacteria, can use atmospheric nitrogen. These bacteria invade the root hair cells and form swellings or nodules.

Legumes are an important economic plant to people, yielding food, energy and domesticated animal fodder. The protein content is higher than grain.

There is a wide range in the different legumes' ability to fix nitrogen, from being completely self-reliant, producing excess or not producing enough for their own needs.

How they do it

The plants and bacteria form a symbiotic relationship. The bacteria is from the *Rhizobium* genus. An important factor to consider is that, unlike other bacteria, nitrogen-fixing *Rhizobium* bacteria require oxygen. It can occur naturally in the soil or be artificially introduced. It is estimated that the world's legumes harvest some 80 million tonnes of nitrogen annually from the air. In factories another 50 to 60 million tonnes is produced annually.

The importance of nitrogen

Nitrogen is a major plant nutrient. It is used by vegetables and crops: for leafy growth, seed production and as an important part of chlorophyll and proteins (proteins are 16% nitrogen). The air is about 78% nitrogen. However, nitrogen gas cannot be used directly by plants and is usually converted to ammonia or nitrates.

Types of nitrogen-fixing bacteria

Yield can be increased by 50% if the right inoculant is used. There are a number of *Rhizobium* species that fix nitrogen. They are usually host-specific and careful matching is needed to get the best results.

Alfalfa group *Rhizobium melitoti*
Clover *R. titoli*
Pea and Vetch *R. leguminosarum*
Bean *R. phaseoli*
Lupin *R. lupini*

Applying bacteria inoculant

The soil where nitrogen-fixing bacteria and their host plants have been living can be used to infect new seedlings. However, there is a risk of introducing unwanted plant seed and disease.

The bacteria inoculant can be purchased, either as a fine peat-based powder for coating seed or blackish granules for sprinkling into seed drills.

Seed treatment

Seed is treated just before it is planted to avoid exposure to high temperature and strong sunlight. Both of these elements are harmful to the bacteria. Thirty grams of inoculant could cover 2 to 4.5 kilograms of seed.

Moisten the inoculant by making a slurry in a bowl. Half-a-kilo of seed will take one teaspoon of water. Add a little molasses or sugar which will feed the bacteria and stick to the seed. The two mixes can be blended together and the seed planted.

Nitrogen soil levels

A little nitrogen will help the nitrogen-fixing bacteria. Five hundred grams of nitrogen fertiliser spread over 100 square metres is recommended, but too much added nitrogen makes the nitrogen-fixing bacteria lazy and inoperative.

How to use nitrogen-fixing bacteria

In a permaculture designed system, nitrogen-fixing bacteria can be used in a number of ways, including crop rotation, green manuring, companion planting, nursery and pioneer plantings, alley cropping and agro forestry for windbreaks.

Crop rotation: growing the same crop or vegetable in a given area is avoided. Different crops are rotated over a number of years (3 to 7). One of these plantings is usually a legume which causes an increase in the nitrogen level of the soil.

Green manuring: is part of (or separate from) crop rotation. The plant can be a legume or a grass. It is chopped into the soil's surface to increase organic matter content in the soil. This leads to humus production and improved structure and fertility.

Companion planting: it is generally thought that when legumes are planted next to other plants there is a nitrogen boost to them. Unfortunately, it seems that this is not the general rule. However, some legumes are known to excrete some 100 to 120 kilograms per acre (0.4 ha) into the soil during their growth. This free nitrogen can be used by other plants growing nearby.

Nursery and pioneer planting: shrubs and small vines are often used as nursery or pioneer plantings. In many cases they are legumes. During the establishment phase of an orchard the productive trees can be interplanted with leguminous shrubs and trees.

Alley cropping: crops or vegetables are grown between rows of leguminous shrubs or small trees. The distance between the tree belts depends on the height they grow to and the slope of the land. The tree or hedge rows are grown on contour lines.

Windbreaks: legumes are usually quick-growing and are thus ideal in windbreaks.

Non-legume nitrogen-fixing plants

There are a number of non-legume, nitrogen-fixing plants. They include useful plants such as *Alder, Casuarina, Ceanothus* and *Elaeagnus.*

Heavily clipped *Alder* can form impenetrable walls or wind breaks, produce firewood and is tolerant of waterlogged soils.

Ceanothus spp. are attractive lilac/blue-flowered shrubs. Some species have edible parts (e.g. Jersey tea) and most are a great pollen source for bees.

Casuarina spp. are quick-growing trees with needle-like leaves which help to break the wind. They also produce timber and firewood.

Many *Elaeagnus* species have edible berries (like olives).

Conclusion

Some things are worth getting into a fix about. One of these is nitrogen-fixing by plants in association with bacteria. It is the natural way that atmospheric nitrogen is made available to plants for their growth and health. An increase in productivity would result at little or no cost. An example of permaculture design and practice in action.

From Soil To Plants

Gardening is usually equated with being a pleasurable pastime, where we can escape from the pressures of modern living. It allows us the chance to get a little closer to nature and ourselves.

Permaculture gardeners also produce food whilst practising this art of quiet contemplation. However, the business of getting back to nature and quiet contemplation is, more often than not, a desperate and anxious struggle to keep plants growing in our poor soils and harsh climatic conditions.

From sand to soil

If we want to grow anything, then it may be important to make some changes to the soil so that nutrients are more available for plants.

A suitable framework is required for the establishment of a soil that will be able to pass nutrients, when present, to the plants, rather than just let them leach away or become bound in the soil. From here it is important to feed the plants suitable organic fertilisers, that are themselves not too limy.

Avoidance of sheep and chicken manures, unless composted, is important. While they both contain a good source of food, they tend to inhibit the intake of nutrients by plants in limy soils, unless used in large quantities.

A manure that I have had much success with recently is pigeon manure, a well balanced mix of macro and micro elements in a form readily taken up by plants. It also tends to be full of feathers that provide the building blocks for humic acids, important ingredients themselves in the development of humus - that all treasured ingredient of good gardens anywhere.

To mulch or not to mulch, is that the question?

All permaculture people know the values of mulching. In fact, it is a strong ritual that brings many permaculturists together (any excuse for a party!). But is it really as necessary as we think?

While I feel extremely incomplete at the thought of having bare ground in any garden that I am associated with, I am starting to come across evidence that suggests mulching may not always be necessary.

Whatever your inclination is with regards to mulching, it is important as practitioners of permaculture that we recognise, as designers, that we are imitating nature as much as possible. It seems that we are applying an exotic concept to our conditions before thinking about developing a strategy drawn from observations of our unique environment.

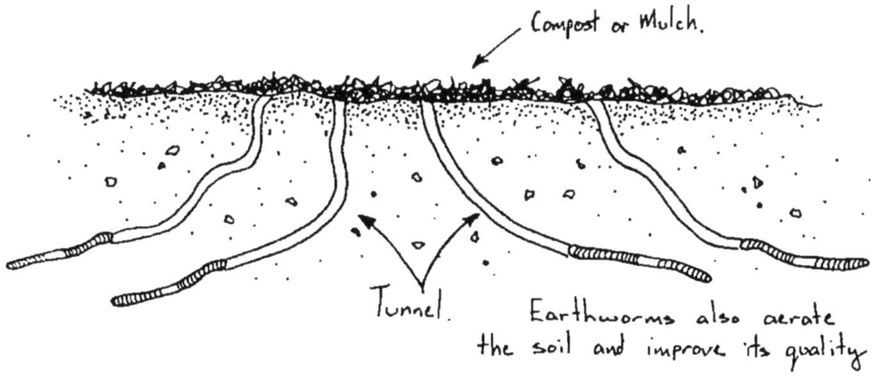

With heavy mulching it is hoped that water use can be decreased. But heavily mulched beds in summer stay reasonably moist as long as they are watered frequently. This moist carpet effectively acts as a perfect shelter for small creatures which means it has the potential to be a perfect place for the 'overwintering' of pests. As a result, the following year numbers can begin at a much higher base level, leading to increased pressures to raise young seedlings to a point where they may be able to withstand high pest numbers.

Maybe the term 'overwintering' needs to be suitably modified and called 'oversummering'! Is it such a bad idea to consider laying the ground fallow over the summer months? It's probably worth thinking about.

From soil to plants

Gardening and growing plants is a very empowering process. Organic gardening the permaculture way seems to empower people in even greater magnitudes. With empowerment comes confidence, and with confidence comes an increase in risk taking. This risk taking can lead to disaster, or even greater discovery.

There are a lot of plants out there that have been forgotten and others with uses that are yet to be discovered. This is an exciting field that I enjoy studying; however, it is no accident that the plants that we have been cultivating conventionally for a while now are generally superior.

Rye and oats began their lives as weeds in cultivated wheat, and in the Americas, tomatoes did the same in corn! The problem arises when this genetic diversity within the varieties is minimised or encouraged for the wrong reasons like enhanced shelf-life or uniform size.

Vegetable growing is just one part of the permaculture plan. The development of perennial systems is a very important aspect too, but, unfortunately, much of our food comes from annual food production.

There are advocates that argue that a switch to perennial food producing systems is plausible but I think we need to have both.

As adventurous permaculturalists, often after growing something once or twice and succeeding or failing, we look for new plants to grow. Different species and different varieties keep our interest going and are lots of fun.

We need to develop suitable seed banks of our more common foods so that we have a reliable source of plant material. With a little practice and refinement it is possible to produce a situation that lessens the dependence on shop-bought food while still having a relatively low maintenance garden.

Ironically, it is not uncommon to see a wonderful 'permaculture garden' abundant in growth, beautiful to look at, but with very little to eat from it. Not that it should be expected that a garden provide all our food needs, but we can make it more productive.

Much to do About Mulching - a Permaculture Perspective

Introduction

One common activity in permaculture gardens is mulching. Many claims are made of it. Some are misleading and it should be understood that it is not a universal solution to the perceived problems of gardening.

However, there are many benefits to be gained by mulching, including those to do with the reduction in the use of energy and the production of wastes.

This article will briefly look at the where, why, and how of mulching and should enable the reader to move into mulching in a confident manner. Much of the information in the article is gained from practical experience over a number of years: what some people would call the 'doings of permaculture'.

Mulching in design

Mulching takes place within a permaculture designed system; one that is designed to be sustainable. It can be a method, technique or a strategy depending on its function. For example, spot mulching fruit trees is a technique, while sheet mulching a garden bed to improve yield would be a strategy.

Two important steps precede the actual construction of the mulch; namely, an analysis of the element to be mulched and the placement within the design system.

Element analysis: annual herb (weed) control

For any analysis a list of needs, products and/or functions is required. The following table would show, for example, that a sheet mulch to cover a large area is required to satisfy the needs and intended products of the analysis.

Needs	Products
prevention of regrowth	organic matter
removal of existing growth	useable space
exclusion of light	
prevention of invasion	

A good sheet mulch provides organic matter and space which then becomes available to other plants. Given that the mulch bed is in an appropriate position you can plant a range of vegetables and herbs which could, in turn, replace the weeds that always seem to come up at the wrong time. An element analysis will help you to understand that the needs of one element can be met by the products and/or function of another. Sheet mulching becomes a strategy which increases the productivity of your land.

Reasons for mulching

The application of mulch material to the soil's surface has many benefits. For example, a spot mulch on fruit trees provides both effective weed control in a small area and organic matter to the growing tree. A larger sheet mulch in a garden bed permits a variety of plants to be grown such as useable vegetables and herbs. Again, weeds are controlled.

Sheet mulches also produce a well-structured, fertile top soil which makes nutrients easily available to plants and attracts earthworms. All that is needed are materials such as straw or sawdust and compost.

The season determines the time

The amount of available organic material for the mulch often determines the time to do it. A permanent cover crop will require little mulching whereas seedlings of poor health or those grown in unhealthy soil may need a mulch for health rather than for seasonal reasons.

In winter, a finished compost which is rich in humus can be applied as a mulch or dressing. Rain will not generally leach the nutrients from the compost if the soil has good drainage. Soil which contains about 10% of humus is considered optimum for a well-structured, fertile soil. Some soils contain less than 1% of humus and the practice of tillage and the hot climate are not conducive to humus formation. An adverse effect of applying a dry mulch in winter is to produce a habitat for snails, slugs and slaters.

The onset of the hot, dry summer is the best time to put on an adequate covering of dry mulch. This will lower the soil temperature, reduce soil moisture losses, reduce unwanted plant growth and prevent wind erosion. Watering needs to be sufficient to penetrate the mulch but not too much that the mulch is broken down too quickly. This is where a trickle system is ideal for direct watering of plants.

Sunscreens and windbreaks may need to be considered. Plants, animals and people are placed under stress when exposed to high energy levels. Cooling systems have to work harder and a great amount of water is needed to avoid structural damage and desiccation. Productivity drops dramatically as a large portion of the system's energy is used in cooling rather than producing leaf, flower or fruit. Ideally the sunscreen and/or windbreak should be productive and thus have more than one function.

How to mulch successfully

I would advise any person mulching for the first time to seek a model to look at. You don't have to find someone else's garden bed to do this. In a natural system such as a forest mulching occurs as the rain of organic matter falls to the ground. Layers are built up and soon the cycle of decomposition into plant nutrients occurs. The bottom leaves are broken down by both large and small living organisms. It is this continual process that enables the wheel of life to continue to turn. The different types of mulching all have one thing in common: they are constructed in layers as you would see in the natural model.

Mulching methods

Spot mulching

A dressing of high nitrogen material such as blood and bone is applied to the area to be mulched. After watering this in, place old carpet or some other membrane covering over the area (however, don't use plastic). If a plant is to be placed in the centre of the area, cut a slit in the carpet or material so that the plant roots will be able to penetrate into the soil below.

A mulch of dry, breakdown-resistant material is then placed on the carpet. If watering is needed, a trickle dripper can be placed alongside the slit under the mulch. You may have to use a brick, stone or small log to hold the spot mulch in place.

One limitation of spot mulches is that the possible invasion of rampant plant species can occur. Spot mulches are only used to get the tree established; they are pioneer type mulches, here today, gone tomorrow.

Sheet mulch

This type is constructed in much the same way as the spot mulch. The main difference is that it covers a larger area set between permanent boundaries. Trees and shrubs can be included in this area. The membrane layer can be omitted if you want lots of living organic matter to grow. Membrane layers are usually used to prevent weeds from growing, becoming dominant and out-competing other plants for nutrients, space and water.

Sheet mulch garden bed

The function of this strategy is to transform an unproductive space, usually having unproductive plants, into a productive vegetable, herb or flower garden in a very short time.

The shape of your garden can vary from squares to rectangles, to circles and to mandala gardens. Its construction is started in a similar manner as the spot and sheet mulch methods. An extra layer of organic matter is placed either above or below the membrane depending on the need. For example, below the material for perennial plants, above for annuals.

The organic matter needs to be reasonably thick in the first year as this supplies the plants with all essential nutrients. Vegetables and herbs can be planted as soon as the mulch garden bed is finished.

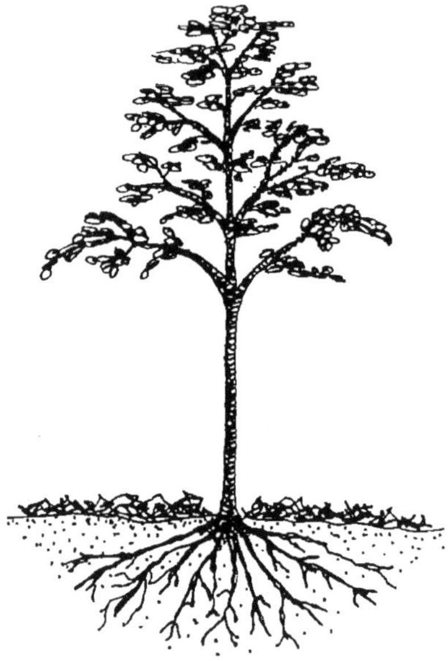

Again, trickle watering systems are ideal while overhead and flood irrigation is a wasteful use of water and should be discouraged as much as possible. Eventually, your established systems of using rainfall in soils with a high humus content can replace high energy systems such as reticulation.

Conclusion

Mulching is more than just a method or technique; it can become a strategy. Sheet mulch garden beds can also have social functions such as you taking responsibility for providing quality, pesticide-free food for your family, recycling household waste, a healthier lifestyle, and co-operation between family members and between a number of families. Mulching needs to be done well to be successful. Invite all of your family and friends to a mulch-in. Household wastes and garden jobs will disappear and be reborn as healthy food.

Mulching

There are serious limitations to using conventional sheet mulch in areas receiving less than 1200 mm/year rainfall which also have less than about 50% relative humidity. Before we get too reactionary however, let us understand the context in which the great sheet mulch myth became so rampant and thereby avoid tossing out the baby with the bathwater.

Many of the early permaculture techniques and practices evolved out of meeting a number of requirements that were commonplace among the 'alternative people' of the southern (higher rainfall) suburbs in the late 70's to the early 80's. By the time permaculture started, many people had embraced recycling and were very conversant with the wasted energy of maintaining lawns, were anti-logging and keen to produce their own food at their houses. These heroic recycling home-gardeners and proto-permaculturists, who felt passionately that they 'couldn't possibly kill anything, had to grow everything in one small garden plot and had to save everything', were surrounded by forests of weeds, vast quantities of newspaper, plastic containers, and jars and bottles that filled the verandah and shed (or both), and had no lawnmowers. Into this ripe and ready ferment stepped Bill Mollison, and after hearing the urban angst said "cover that lawn with those there newspapers and old carpets, use the plastic bottles for seedlings and with the abundance from your new world permaculture garden fill the bottles with wines and your jars with preserves (brandied of course)". Well as they say, "the rest is history" and we can only wonder at what would have become of us all if Mollison hadn't cross-linked all those loose ends for us!

Many years later and well away from the regions and context that spawned these practices it has been discovered that the thick organic layer of sheet mulch stops the water and light rains getting into the soil and 'wicks' it back into the dryer air typical of most Australian climates. This effect is the opposite of what is actually claimed to be the benefit of sheet mulch and any rational gardener would give permaculture the flick if they saw this demonstrated as the way to go.

There is still quite a bit of merit in using this technique for suppressing weeds and killing lawn, but be warned that it will definitely inhibit soil development, starve plants of soil moisture and increase water bills beyond your wildest imagination in areas of low rainfall.

OK, so now we can focus on mulch as a concept, rather than a specific style or medium, and look at what we want to achieve by the use of mulch. Primarily, moisture retention, shading, cooling or warming, improving soil fertility, erosion control, suppressing weeds and fire, and trapping seeds are some key objectives. Try not to get caught in dogmatic and prescriptive thinking and 'import' straw from the wheatbelt to some area many hundreds of kilometres away just so you can kid yourself that you're doing it 'correctly'. This is not permaculture, is a waste of money and you are firmly back on the fossil fuelled treadmill of conventional agriculture.

Each locality will generally have some form of material that can be used as mulch to secure these objectives. Careful observation will lead you to discovering how nature is achieving these characteristics in your locality. The most dynamic form of mulch is a carpet of 'living mulch' such as groundcover plants (for example, purslane), that are both water-thrifty and dynamic accumulators of nutrients.

Moisture retention is achieved by allowing the easy passage of moisture into the soil and inhibiting the dry air, wind or water-hungry plants from stealing it. Keeping organic materials below the soil surface avoids 'wicking' soil moisture off into the air.

Most organic sheet mulches act as surface water storage reservoirs prone to evaporation losses, when the aim is to store moisture in the soil itself, which is where organic materials are needed anyway. Watering the soil rather than the plants encourages deeper, cooler, water-thrifty roots, while making the plant less prone to drought and overheating. By using impervious, non-porous, hydrophobic materials like hard rocks, stones, pebbles, slate fragments or sand as mulch, water can easily pass into the soil and soak in.

Shading and cooling can be achieved by shielding the soil from hot winds and minimising its sun exposure during the hot months by covering it with light coloured, reflective, low-thermal mass materials such as white quartz pebbles/stones, coral fragments, shells (even snail shells), bones or a thin scattering of fresh young leafy branch trimmings (as they bleach to light cream colour as soon as they dry), palm fronds (top or face down so they don't 'cup' any water), or a sparse veneer of sheep wool and ash.

Warming the soil can occur by keeping it shielded from cold wind and providing it with a sunny aspect during the cold period, leaving it bare or using dark coloured high-thermal mass materials such as a single layer of dark stones/pebbles, gravel or small charcoal fragments, old cast iron car parts, thin layer of bitumen or old road pieces.

Any hard, dark stones or concrete fragments spread in a single layer with intimate contact with the soil will help to transmit warmth.

Soil fertility, erosion control and seed entrapment are all markedly improved by the above mulching strategies, while each specific technique will give rise to outcomes that are unique to each technique. For instance, old bones may attract unwanted attention from dogs but the use of their own dried, sun-bleached manure would be a good substitute for bones and yield more readily available phosphorus, while acting as a deterrent to their digging.

Weed suppression immediately around most young plants to a minimum of about 0.5 m diameter is often desirable and to this extent there is a role for limited sheet mulch. Another technique that works well is to use a 1 cm thick layer of clay over the newspaper instead of the straw or woodchip, both of which can still host incoming seeds.

When the clay is dampened, packed down smooth and then dried, it offers a very unattractive hard packed surface to most incoming seeds even if it gets wet and soft. It is as useless at allowing water transmission into the soil as sheet mulch, and as such should not extend over more area than is necessary.

Generally most plants only require this sort of protection and support for their first couple of years, whereupon they start to take control of their immediate surroundings.

Fire suppression is somewhat more involved and to a large extent applied to groups of plants rather than individual plants. Primarily there is a fire risk to plants that are surrounded by a fuel load - comprising dead grasses, twigs, leaves, woodchip, straw, branches, logs and/or living flammable plants.

Immediately one can see that there is a major fire risk from most conventional sheet mulches or overgrown self-mulched systems. The rock, stone, pebble, clay, shale, shell, coral and sand mulches suddenly have an extra advantage over more commonly used materials in providing individual plant protection in the early years but would not offer long term protection once the plants start self mulching.

The one form of mulch that can reasonably be expected to meet all of the above criteria, but which requires more extensive knowledge of the plant kingdom is the use of living mulch. There are many small groundcover plants that are water-thrifty, fire retardant, dynamic accumulators of nutrients, shady, quick-growing and, once established, inhibit or suppress other less-desirable plants from occupying that niche.

A number of these are also food or fodder and/or medicine plants as well. The other advantage is that if you choose the plants that have fire suppressant qualities as well as being good nurse/mulch plants, there is often long-term fire protection from these species if they end up spreading around between the mature trees and shrubs filling the intervening spaces.

So once again two of the primary maxims of permaculture that 'diversity gives rise to stability' and 'every function is carried out by a number of elements while every element provides a number of functions' are found to be more substantially fulfilled by the use of living mulch as distinct from other mulch types.

The area immediately under the 'drip-line', for example, receives considerable 'canopy runoff' and is often not impeded by a thick layer of mulch. The area between plants which is a lot more bare receives and easily absorbs all available moisture. Thinner leafed or more open canopy plants often conduct moisture down their limbs and trunks into the soil at their trunk's base where mulch is less compact.

It is the many subtle interactions that take place in nature that we must observe if we are to become 'practicing permies' and not book readers.

On Brushwood Composting

Most people must have noticed that shrubbery and tree prunings left in heaps to either cart away or to burn after the winter produce an abundant and healthy growth of weeds and grass around the perimeter and through the heap. So why waste this valuable material by burning it or waste valuable energy by taking it to the tip? Prunings can be made into the "Rolls Royce" of compost.

Twigs up to pencil thickness break down quite well in twelve months or less if set up in a proper compost heap, with a little care. The thicker branches can be shredded and used as well, provided it is living and green.

The whole secret of the shredding is that it is sliced longways in slivers not much more than 1 mm in thickness so that bacteria have the maximum surface area to work on decomposing the lignin (woody substance) in the material.

The next most important thing is moisture content. This can be achieved by soaking the material with a fine spray or by soaking it in, for example, an old bathtub or other suitable container. Next comes the setting up of the heap. The best results are obtained by creating a heap of at least 8 m^3. It can be set up in any shape, as long as it is 8 m^3. Personally I prefer a rounded heap.

As the heap is set up layer by layer, jump on it to compact it. Put down a layer of approximately 20 cm, stomp on it until there is not much give left and repeat the process until the heap is complete, ensuring that each layer is soaking wet.

Now you finish the heap by spreading a 2 cm layer of sand, or garden soil, over the lot then cover that with old Hessian sacks, honan squares, branches of pine trees, bracken or similar.

In hot weather keep the outside moist by spraying it with a fine spray or put a sprinkler on top. It is crucial that the heap does not dry out. After 8 to 12 months turn the heap by raking the sides and top down to become the core as you turn it inside out.

As you are turning it and setting up, give it a fine spray of water. This time it is set up loosely and again covered as before. The heap is ready for use 1 month later, and you will have the finest soil food in nature!

Now to use this "green gold". For vegetables, spread in the first year a 5 cm thickness of compost on top of the soil and cover with a 10 cm layer of mulch. For mulch use straw, or whatever you can lay your hands on. Pine needles are good as long as the mulch is removed after the summer months and the compost can be lightly forked in or left as it is. It is very important to cover the compost with mulch immediately after spreading it.

Now plant your seedlings into soil underneath the compost by spreading open the compost and mulch. Replace the compost and mulch around the seedling, giving it room to grow - do plant in the soil. The compost feeds the soil and the soil feeds the plant.

If you set up a 100 m^2 garden completely covered by compost and mulches you will find that you don't have to water your plants, if you can keep them shaded from the hot sun. As the plants get bigger you can remove the shade. Wilting will occur in the hottest part of the day but the plants will stand healthy and vibrant when the sun moves over in the sky.

After the first year you will need 2 cm of compost yearly and the 10 cm mulch on top.

I will now give quantities of compost for use yearly for grassland, grain and fruit crops and recipes for seedraising and potting mixtures.

Grassland

(At the time of planting)

Poor soils: 30 tonne/hectare

Medium soils: 20 tonne/hectare

Good soils: 10 tonne/hectare

(Thereafter use 2 tonne per hectare yearly)

Grain crops

Poor soils: 120 tonne/hectare

Medium soils: 80 tonne/hectare

Good soils: 40 tonne/hectare

(Repeat yearly)

Fruit crops

Poor soils: 30 tonne/hectare

Medium soils: 20 tonne/hectare

Good soils: 10 tonne/hectare

(Repeat yearly)

This is equivalent to 50 kg per tree at a plantation density of 200 trees/hectare.

Seedraising mixture

One part sand or garden soil.

One part screened brushwood compost.

Potting mixture

One part sand or garden soil.

2.5 parts brushwood compost.

Using brushwood compost you will grow healthy, nourishing crops with a minimum of diseases and pest attacks.

You also save on water, have less work to do (no weeding) and you can go on holiday knowing that when you return your garden will still be there growing happily.

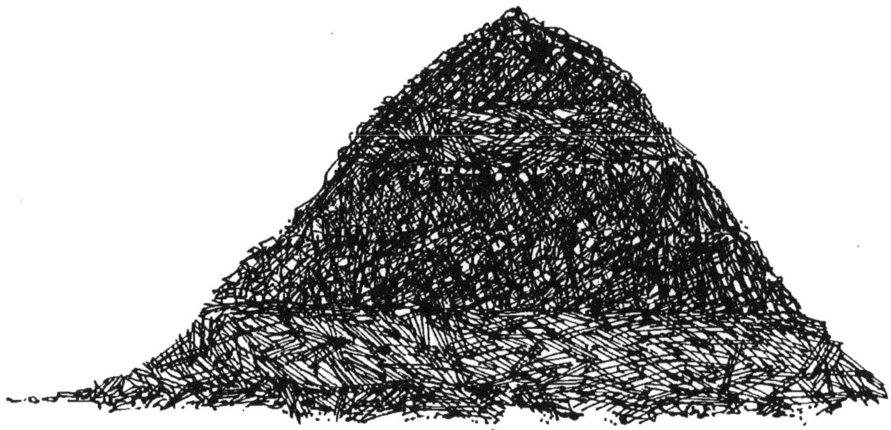

A Natural Alliance Between Permaculture and Integrated Pest Management

As an entomologist and a student of sustainable agriculture, agroecology, natural systems agriculture and now permaculture, I often wonder how necessary are my specialist skills in biological control (using bugs to control bugs). If pests are simply a problem of monoculture (growing large areas of the same thing) then polycultures or mimicking nature is the answer and people like me are redundant. Likewise, if only unhealthy plants are attacked by pests then creating optimum growing conditions is what is necessary and again I am redundant.

Nevertheless, I do argue that an understanding of pest species and their natural enemies is useful for creating balanced systems. And because an agricultural system is by definition an artificially created system, a managed system, it will be less than perfectly balanced.

Pest outbreaks are indicators that the system is out of balance, whereas low population levels of pest species can indicate that the system is healthy. Natural systems do have occasional pest outbreaks, albeit in patches and usually as a response to extreme or unusual weather conditions. Likewise, an agricultural or gardening system can be put out of balance by extreme weather conditions, but also by too much or too little nutrition or water, and particularly by the use of broad spectrum pesticides.

Integrated pest management or IPM has many definitions. The one I use is close to the original definition of a pest management strategy which utilises a range of tools, with highest priority given to those that mimic nature, which encourage complex biological interactions, thus increasing stability, and which are least environmentally disruptive. This approach is similar in my mind to the principles of permaculture.

IPM as I define it has two facets; firstly, it involves the design of a system that is generally resistant to pest outbreaks, and secondly, it entails an appropriate response when pest populations are increasing. Managing pest populations has much in common with the design process in permaculture. It is an interactive process based on the observation of natural and designed systems and involves the mimicking and monitoring of these systems. We can build into our designs aspects that promote resilience but since conditions are always changing and developing, so we will always need to observe the interactions in our gardens or agricultural systems and respond if needed.

Designing a pest resistant system

It is simple logic to realise that if you provide hundreds of hectares of a single plant species, then anything that feeds on that species is likely to develop into a pest. Likewise, if you systematically kill the natural enemies of that pest then the pest will continue to multiply. Therefore one essential element of a pest resistant system is to increase diversity. The permaculture approach to food and fibre production is to mimic the natural forest system thus creating a system which is diverse in both plant species and plant structure.

Generally, a diverse system makes it more difficult for pest species to find suitable host plants (it can also make it more difficult for natural enemies to find host pests), and it provides conditions suitable for generalist predators. Flowers rich in nectar will encourage parasitoids (organisms that develop in or on another organism and kill it in the process - usually these are wasp species) and many other important predators into a system, as many of them feed on nectar as adults and only on pest species as larvae.

Within broadacre farming, diversity can be increased in a variety of ways including tree belts, intercropping, crop rotations, weedy verges, agroforestry, strip cropping, and multi-cropping. Nevertheless, in spite of these measures, pest species may still build up in sufficient numbers to cause economic damage.

Dealing with pest outbreaks

In case of an outbreak the first question to be asked is whether it is causing sufficient damage to require a response.

The presence of a pest is not necessarily an indication of a problem. Since 1945 and the advent of pesticides, it has been a common belief that a healthy system is a system devoid of any pest species and that such a system is both possible and desirable.

However, despite trying to eradicate pests in crops for fifty years through the application of ever-increasing quantities of pesticides, the proportion of crops damaged by insect pests is proportionally similar to the days before pesticides.

As an ecologist it bemuses me that to create insect pest outbreaks for the purposes of study, it is common practice to spray an area with pesticides and then wait for the pest species to build up.

Even if you never use pesticides and have created a diverse system, pest populations will increase from time to time and a response will be needed.

Firstly, pest species tend to be like pioneer plant species in that they reproduce rapidly, usually having many generations in a year and often many offspring at a time.

However, some insect species classified as pests have only a few generations in a year and relatively few offspring. These species are termed low-density pests and the codling moth in apples is an example of such a pest.

Pests and damage

Pest species can be classified by feeding habits: there are sap-suckers such as aphids; leaf-chewers which include cabbage white butterfly larvae; borers which include many weevil species; and stored-products pests such as grain moths.

Pest damage can be classified as either "direct", meaning that the damage is to the part of the plant or crop to be harvested, or as "indirect", which refers to damage to a part of the plant that is not to be harvested but which may or may not reduce the harvestable component.

Generally, a system can tolerate much higher levels of indirect pest damage than direct pest damage, and in some instances indirect damage may actually improve the resultant crop (for example, late season defoliation in grapes can improve air circulation and reduce fungal problems).

If you have a pest species building up in your garden or crop you need to ask yourself a number of questions in order to arrive at an appropriate response.

One is whether the pest is really a "pest", and that answer will depend on many factors specific to you and your system. For example, you have to ask whether the damage is detrimental to the harvested produce, is the crop necessary for your survival and is the cost in time and money of a control measure less than the cost of losing your crop.

Another question to ask is whether the infested plant or crop is growing in optimum conditions. Stressed plants often attract pest species or suffer more for the damage that is done; it is therefore a basic pest management tool to create optimal conditions for your plants as much as possible.

It is also often true that after pest numbers build up, so too do the numbers of their natural enemies, and that waiting for the natural enemies to do their work may be an appropriate pest management strategy.

Identification of the pest is an important part of your strategy. What is the species and what do we know about its life-cycle, living conditions and its natural enemies? Useful information about a pest includes the number of generations the pest has in a year or season, what is the pest's preferred host and will it feed on other hosts, does it require an alternative host, is it a native pest or an introduced pest, does it have natural enemies (predators, parasitoids, pathogens, viruses or nematodes) and are they found in your region, what are its optimal micro-climatic conditions for each life stage, where is each stage found (for example, codling moth eggs are laid on leaves or fruit, the larvae burrow into the fruit, the pupa is normally hidden in bark crevices and in some cases in the soil, the moths shelter during the day in the foliage and fly at night), and often an important "weak-link" is how and where does the pest over-winter.

From this information a simple management practice may become evident, such as timing planting to avoid a pest egg-laying period or weather conditions which favour the pest.

Likewise it may become evident that the pest finds the host plant by visual cues and can thus be deterred with a barrier of non-preferred plant species.

Management tools

Management tools or strategies can now be matched with the pest. These tools include cultural, biological, bio-rational and chemical controls.

Cultural control; that is, changes in crop management to discourage the pest or encourage natural enemies, should be given higher value than more environmentally disruptive approaches such as the use of chemicals (whether plant derived or industrially produced).

Cultural controls may include altering row spacing, planting and harvesting times, variety choice (resistant varieties), watering and fertilising practices, and altering or introducing intercropping or trap cropping.

Secondly, biological controls can be considered. Biological control is the deliberate use of biological organisms to control other biological organisms. Generally biological control is divided into classical and augmentative biological control.

Classical biological control involves the introduction of natural enemy species from outside the region with the purpose of neutralising the enemy and keeping the targeted pest below economic damage levels. Augmentative biological control involves the regular addition of insectary-reared natural enemies into the crop or garden system to supplement the local natural enemy populations.

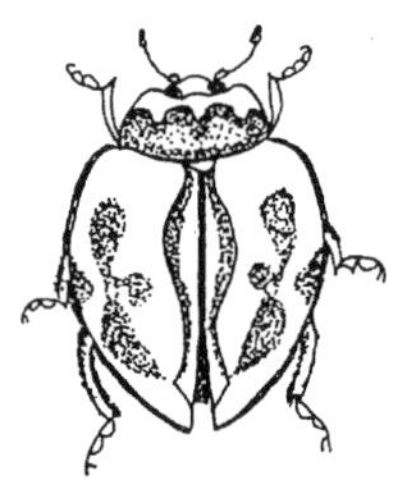

The scarcity of commercial insectaries in Australia makes it difficult for many individuals to utilise biological controls. Over time I expect this situation will improve allowing more emphasis to be placed on this approach.

Classical biological control should only be carried out by professional people as it requires extensive testing of the natural enemy species prior to release to ensure suitability and, more importantly, to eliminate potential environmental impact.

Classical biological control often conjures up memories of cane toads and other such disasters where testing proved to be inadequate. The disasters of biological control have usually been with vertebrate control agents. As vertebrates are generalist feeders (meaning they will eat many different species) it is not surprising that introduced vertebrates have shifted from the targeted pest to other non-targeted organisms. Invertebrates on the other hand are more specific feeders and so tend to be safer to introduce. The ideal classical biological control agents are the parasitoids, which have co-evolved with their hosts and have had to outsmart their host's immune system. These agents are very specific to the host and will rarely develop on other species. Likewise, insect viruses have very narrow host ranges and can often be safely introduced.

Next, bio-rational control measures use tools which utilise some aspect of the pest's behaviour or biology to kill it or interfere with its reproduction. An increasingly popular bio-rational measure is the use of pest pheromones.

Pheromones can either lure the pest to a place where it can be trapped or they can be used to disrupt mating patterns. There are also a range of traps and physical barriers that can be included as bio-rationals. Insect growth regulators are also being widely used in some parts of the world. Insect growth regulators use growth hormones and inhibitors that are specific to pests and in some cases are specific insects which hormonally disrupt the normal lifecycles of the pest.

Some work is also being done on various feeding deterrents. *Bacillus thuringensis* or Bt has been widely used for the control of caterpillars and some beetle larvae, and essentially acts like a pesticide - it is a stomach poison.

Finally, chemicals should always be used as a last resort because they are short-term solutions which disrupt the whole system to a greater extent than other control measures.

Many chemicals are vertebrate poisons - including many of the plant-derived chemicals. And there is always a risk of pests developing resistance to the chemical, and it is this resistance which drives the so called "chemical treadmill", where ever increasing doses of the chemical are needed to produce the same result.

"Transgenic" plants are now coming onto the market. These plants have genes spliced in so that the plant itself produces a pesticide. At present Bt has been spliced into cotton, tomatoes, potatoes, and a range of other plants. Having a plant express pesticides is probably the quickest route to pest resistance and is therefore a particularly dangerous and short-sighted approach.

Information

For common pests the information can be gleaned from gardening and garden pest books, and often friends or neighbours. There are a number of good books which will give the basic information on common pests.

Every region is a little different and the strain or species that you have in your garden may be slightly different from the ones in the books so it is really important to make observations of the effects of your management practices. Even with the less common pests that you may not be able to identify or get the relevant information on, observation and experimentation can go a long way.

See the Solution, Not the Problem

There are many joys in growing your own food organically and one of them is sharing your bounty with friends. Unfortunately, there are some friends that have not been invited to share the bounty but attempt to have their share before the vegetables have even grown. Yes, I am talking about those pesky molluscs - snails and slugs.

There are a few different ways to deter the prime suspects and one of them is probably the most famous and favoured in summer, and that is the beer trap.

Yes, a little for the gardener and a little for the snails. In fact, the best lure is the dregs left over from home brew, because the secret is the yeast in the brew that attracts the slippery little suckers.

But, if you are tee-totallers or don't have any beer to spare, try a little vegemite or any other yeasty product. Just add it to a bit of water in a recycled container and bury it in your mulch so that the opening is level with the ground and the intended victims will follow the smell and drown happily.

The thing to remember is that if your garden has overhead watering, or if it rains, the mixture will dilute quickly so keep it topped up.

One of the most effective defences against slugs, snails and slaters is the orange cup. After you have squeezed yourself a juice, place the skin upside down near new seedlings (that is the dome pointing upwards).

Go out into the garden, first thing the next morning and collect them up, neatly in the cup. This is much easier than hand-picking and best results are definitely early in the morning.

Diatomaceous earth is a barrier far superior to sawdust, lime or ash. Apart from the fact that the other products may imbalance the pH of the soil they are rendered inefficient in wet weather.

Diatomaceous earth is a lot of razor sharp particles which the snails and slugs cannot traverse; be sure to wear a mask when handling this substance as it is not too good for our lungs. Simply make a barrier of this around your garden bed or newly planted seedlings in an established bed.

You can use wormwood from your garden to make a very strong tea and drench the bed with it. This must be repeated every few days until you see a difference. If I make it as a spray, I like to add chilli which makes it even more effective as an all purpose spray.

Another product I have used with effectiveness, most notably with slugs, is Neem oil. You can buy this product with added surfactant ready to be diluted - 10 millilitres to one litre of water for slight infestation or 15 mL to one litre for a heavy dose.

This stuff actually works because it is really very bitter and the pests do not like the taste, therefore they will not be so inclined to munch away at those succulent seedling morsels. If they are persistent enough and continue to feast on your garden, the neem will then interfere with their birth cycle by mutating their young, therefore cutting out future generations of pests.

A close relative to the Neem tree is the White Cedar or the Cape Lilac, a tree which is very common in the suburbs; their leaves are also a molluscicidal. I have read that the leaves of *Gardenia jasminoides* are as well, but I haven't tried that one.

Also don't forget biological control by encouraging birds, small reptiles, such as lizards, and amphibians - that means frogs. And even more control would be ducks, preferably Khaki Campbells (they have the best reputation for eating snails). Although ducks don't scratch up gardens like chooks they do have a tendency to flatten plants so use them to clean up your patch before you plant or as a barrier patrol.

Integrated Fruitfly Management

Background

To date, all attempts to control the Mediterranean Fruit Fly (Medfly) (*Ceratitis capitata*) in the South West by local and state authorities have relied almost entirely on the use of organophosphate pesticides.

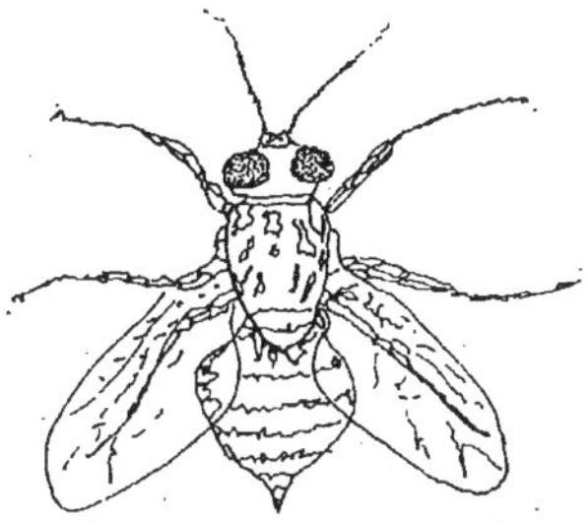

However, in response to mounting community concerns about the use of pesticides on horticultural products, agricultural authorities should accelerate the development of non-chemical, pre-harvest procedures to replace them.

Several other species of fruitfly may also damage fruit but are not considered to be as significant as the Medfly.

The following points and strategies are relevant to both commercial agriculturalists and backyard growers:

i) 100% protection of susceptible soft fruit may only be achieved by repeated heavy chemical application using such toxic organophosphates as Rogor, Lebaycid and Dipterex.

ii) The above chemicals are toxic to non-target beneficial organisms including useful predators and also humans.

iii) It is likely that residues of the above chemicals will be eaten with the fruit, if the fruit itself has been treated.

iv) Non-chemical or minimal chemical management should be a long-term program aimed at use over a wide area which may necessitate discussion and co-operation with neighbours.

v) Methods of integrated management may include physical, biological, cultural and sometimes low-hazard chemical management.

Note: When an inquiry is directed at the authorities or most 'experts' regarding fruitfly control by the home or commercial fruit grower, there is the tendency to point straight to a chemical-intensive solution, thus perpetuating a vicious circle. To escape from chemical dependence, as with other dependencies, the need for patience and the ability to observe and modify responses accordingly cannot be over-emphasised.

It is worth noting that there is no guarantee that the methods outlined will be totally effective as research into such strategies is ongoing and much of such a program relies on a property owner learning about an individual situation and modifying it under local conditions and in cooperation with neighbours.

Monitoring

Monitoring is a critical process independent of whether a totally chemical dependent regime, an integrated approach or a fully organic strategy is to be followed. Monitoring may determine the timing of a damaging population increase and the direction from which an external (immigrant) source of infestation arises.

Energy, time, materials and money can be wasted or misdirected if the timing of any management strategy does not coincide with the presence of the pest or a build up of its population to a damaging level, or if an infestation is largely immigrant in nature.

Visual inspection of fruit trees may not always indicate the obvious or imminent presence of economically significant numbers of Mediterranean fruit flies. They will often be in other species of trees perhaps feeding on fungal spores or sweet secretions from other insects or on nectar from blossoms. They may not be in your trees at all, due to effective in-house pest management, but waiting in your neighbours' orchard or garden.

A most effective method of determining presence or activity level is to trap, count and chart. Counting should be carried out on a daily basis (if time permits) or one day per week at the very least. Dry-baited transparent male pheromone traps are the most effective for this purpose.

Assuming that an equal representation of both sexes is likely to occur in a wild population, a rapid increase in captured males signals the imminent occurrence of egg laying by females in fruit on the tree. Assuming that a female mates on the day she hatches, with ample food and optimal temperatures, at least seven days must pass before egg laying occurs. If conditions are not optimal this period will be extended.

Strategic placement of traps combined with accurate recording of location, dates, wind speed and direction which correlate with a significant catch can all be extremely useful in determining the pattern of a developing infestation. In an area of definite immigrant activity, placement of traps at different heights around the orchard perimeter may be useful in planning for the maximum height required for placement of any physical barriers (e.g. vertical netting).

Monitoring, if religiously carried out over several seasons, will often lead to a high degree of predictability of outbreaks. The availability of multi-variant computer analysis facilities should not be overlooked in correlating such variables as time of year, wind direction and other weather conditions. Such resources may well be shared between a group of neighbouring fruit growers.

Management

Physical management

i) Collect all fallen fruit and dispose of it by immersion in a large tub or bucket containing soapy water placed permanently near the trees, or effectively compost or mulch. Grazing sheep, goats, pigs, geese or chickens at fruit fall time is often an effective and economical option.

ii) Band the trunk of trees with sticky non-drying pest adhesive or tight-fitting teflon tape in early summer to exclude ants from the tree and thereby reduce the amount of sugar and subsequent black sooty mould available from aphids and scale etc, which are spread and farmed in the trees by ants. The sugary secretions and the mould are known fruitfly foods. (Note: Wait until well after annual mass hatching of predatory ladybirds has occurred if possible. This may vary depending on local and annual climate variations but will generally occur around September/October).

Banding should be placed as far up the trunk as possible to reduce the chance of dirt, twigs, grass or other debris bridging the sticky band.

Sticky bands should be at least 3 centimetres high and approximately 3 millimetres thick. Tar paper, thin felt, rubber, plastic tapes or other bases are recommended to be used between the sticky material and the tree trunk to avoid potential damage to the bark. This is despite claims by some that direct application is safe.

Experiment on several trees if you have any doubts. Always ensure that any base used is in total contact with the trunk around the entire circumference! All other ant access routes into the tree should be eliminated by skirt pruning and keeping grasses down.

iii) Treat trees from time to time with soapy water and white oil for scale and/or aphids. This is most effective following banding. Derris dust may sometimes be appropriate as long as it is used well away from water courses.

iv) Regularly during the fruiting season consider lightly scarifying the soil around the base of the trees to expose some of the larvae, pupae and emerging young adults to dry heat, light and predatory creatures such as ants. (Supplies of sugar water and pieces of timber on the ground for ant nesting may increase the attraction of ants to these trees). This approach may not be feasible where heavy mulching has been used. Foraging poultry may also be used to help control Medfly in the soil.

v) Use traps and baits. These can be used in large numbers and in varying forms:

- Buy or make sticky traps designed to attract males (using sex pheromones) or females (using yeast or protein attractants).
- Make or purchase traps known to be effective in fruitfly capture. Try a range of baits to satisfy a range of preferences; eg for female Medfly try vegemite/yeast, urine, cloudy ammonia and sugar, stale beer.

Hang these amongst the fruits and rebait when necessary. Dry baits are preferred as identification of the target organism is made easier. For male Medfly, species specific synthesised sex pheromones are available.

Male Medflies are identified by the presence of clubbed antennae. Females possess a pair of straight bristle-like antennae lacking the clubbed ends. Expertise in identification is assisted by the use of a 10x hand lens. Only the female flies sting the fruit.

vi) Immigrant barriers. In a well-designed monitoring program it may be evident that all or most of the fruitfly population is arising from a neighbouring property or from a specific area or direction and that a height ceiling for flight exists. In such a case vertical insect mesh barriers may be effective in blocking the flight path. If monitoring has not determined the height ceiling, barriers approximately twice the height of the fruit trees should be erected. Traps are best placed outside the barrier at points where it angles, and/or large funnel shaped traps may be made in the barrier mesh.

The presence of dense vegetation predator habitat barriers will contribute to a similar effect. Note that if the main source of infestation is from within the production area, such a technique is unlikely to have a significant impact on numbers.

vii) Heat or cold treatment techniques for fruit have recently been developed which will kill larvae in the fruit after picking. 100% success is being gained under certain conditions including the area of overseas export.

Cultural management

i) Attempt to grow fruit which will ripen well before the fruitfly breeding season. Ask your specialist nursery staff or other experts for the best varieties.

ii) Pick fruit at the earliest possible stage before stinging and ripen it inside or during transport if feasible.

iii) Pick ripe fruit and store in refrigerator or bottle. Accept a certain degree of damage knowing that benefits exist from not handling or consuming unknown volumes of chemicals.

iv) Plant a diverse garden of many species of quick growing shrubs, trees and herbs around and amongst your fruit trees. This will over the long term provide habitat for predators and give some visual camouflage for fruit from roaming flies (which rely on sight as well as smell) and help to deflect breezes which may otherwise carry the scent of ripening fruit downwind.

Albizzia lophantha, *Acacia decurrens*, *A. mearnsii* and *A. baileyana* are known quick-growing ladybird attractants and egg-laying sites.

v) Some aromatic plants may even help disguise or mask the odours favoured by the fruitfly. Experience will dictate here. Try scented geraniums, wormwood, southernwood (Lad's love) and others.

vi) The removal from the area of fruit trees such as loquats which will act as intermediate breeding hosts and food sources is strongly recommended.

vii) Discuss with your neighbours the long-term strategy of reducing chemical dependence and ways in which change can be afforded by co-operative means.

Biological management

Ants are the most convenient approach to biological management with some species useful for the predation of larvae and pupae. Poultry and native ground-feeding birds will also contribute to effective control of pupae and larvae in soil beneath fruit trees. Other effective predators would include flying birds (such as swallows) dragonflies, bats and honeyeaters.

The potential for biological control by increasing local system diversity must be appreciated. Such organisms as pathogenic nematodes (to eat the larvae), protozoa, wasps, fungus, bacteria and other diseases may be commercially available in the future.

The use of sterile male releases has occurred as a component of management and the expertise and technology is available from local authorities and private enterprise. We should always be requesting that our authorities move towards adopting such strategies.

Chemical management

Should any non-selective chemical be used for domestic fruit protection I would only consider using low-hazard materials inside traps or possibly as splash baits to give a quicker degree of kill to specific insects which may otherwise escape.

Many chemicals currently registered for control of certain insects are known to affect the behaviour of a range of organisms, including those which would normally help control the target species. Lebaycid, the most commonly used chemical treatment, is highly toxic to birds and should not be used in situations where such exposure will occur, according to the registered label. Maldison is highly toxic to many insects including bees. An alternative splash bait is a mixture of 3 parts boric acid to 1 part borax used at 5% in a Flavex solution splashed on leaves and neighbouring trees.

Resistance to a range of chemicals in insects is widespread with a number of cases documented where the chemicals have even caused such bizarre results as increased reproductive rates in the pests.

The issue of chemical use is highly relevant to commercial, domestic and export production as markets demand a "cleaner" range of produce.

The above strategies are not guaranteed in any way to eradicate completely the insect targeted. They are part of an integrated management philosophy. Where the market demands 100% "sterile" produce in terms of pests or blemishes it must expect a residue of pesticides to be contained within that produce under current conditions.

Trapping and baiting

Traditional attempts by the public to trap Mediterranean fruitfly has usually revolved around the use of various liquid baits containing sugars, protein or derivatives of these. Such baits as stale urine, cloudy ammonia, vegemite solution, stale beer or sweet wine have been used with varying success.

The main problems with such baits are that they often catch a range of creatures (including the ferment fly - *Drosophila spp.* often incorrectly identified as the fruitfly) which end up in a dark unrecognisable soup in the bottom of the trap. This is certainly not conducive to keeping any definite tally of when the real Medfly is active.

Types of traps have often ranged from the open unlidded jar to the more sophisticated plastic PET drink bottle with several small holes in the sides. Of course these traps will work to some degree and may be sufficient to keep numbers down in low population areas. They are certainly cheap to construct and maintain.

Home made traps can be cheap and use recycled materials. They may be time consuming to construct in large numbers however. The one kilogram transparent plastic honey buckets with the orange/yellow lid are our favourite. We create two one way funnel entrances on opposite sides but at slightly different heights, with a bait and killing agent inside. We have also had success with a range of sticky traps using both colours and volatile baits as attractants.

Baits or lures

Male: the male pheromone Trimedlure is available in the form of a slow release polymer plug which is highly effective for several months. This is a dry bait.

We suggest for purely economic reasons that each plug as supplied may be cut into either two or even three segments without losing significant effectiveness. Similar products under different trade names are also available through other outlets here and abroad.

Female: a range of liquid protein hydrolysates are available commercially (e.g. Flavex). The use of a saturated solution of any yeast (e.g. Promite/Vegemite) extract will do much the same job.

A mixture of cloudy ammonia, vanilla essence, raw sugar and water is also effective. To convert a wet bait into a dry bait add a neutral wetting agent (e.g. a soil wetting agent) to the liquid and soak caneite cubes in it until saturation occurs.

Drain the cubes and sun or oven dry. These can then be stored in large numbers in the fridge in sealed plastic bags and used to replace old baits. Replacement of female lures may be necessary every two weeks (observation will determine this).

Killing agent for use with dry baits: the commercially available Dryacide dust (locally mined and manufactured) is a non-toxic and low-hazard amorphous silica dust registered for protection of stored grain against insect attack.

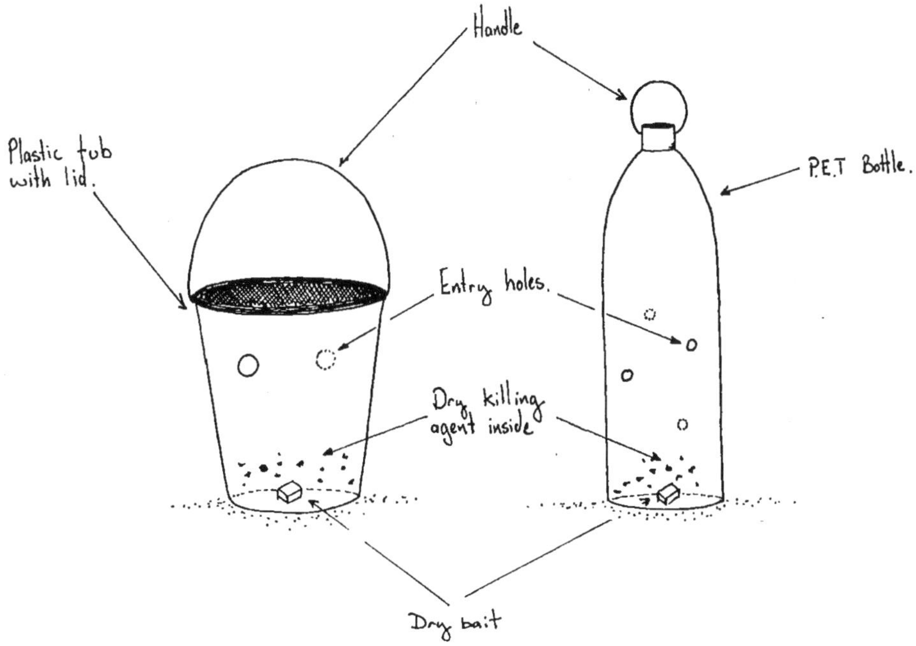

Inside a trap it appears to be non-repellent to male and female fruit flies (unlike malathion) which increases the likelihood of a high capture rate of recently mated females and searching males. Less than one quarter of a teaspoon of Dryacide in each trap will last for months and will lead to rapid desiccation of any insect which comes into contact with it.

Even fruitflies which enter a trap and escape with Dryacide on their bodies appear to either fall to the ground or die soon after, usually without egg deposition.

Trap Placement to Maximise Mating Disruption: a large number of male traps should be placed within the area to be protected. The greater the number the better.

An area continuously saturated with male attractant will not only tend to keep them from finding the real females but will also help reduce the total numbers by mass trapping.

Female traps are best kept away from the area to be protected to maximise mating disruption - with some traps touching the borders of the fruit production areas. This technique will work best when most of the threat is from an immigrant source.

Trapping Out to Reduce Population: as per above but if infestation is primarily from an internal source, the separation of traps may not be as effective in disrupting mating.

In this case, mass trapping using as many female traps as possible (wet or dry) is the preferred option. Some dry traps should be used to allow for monitoring of population movement.

Poison Baits: Malathion mixed in a liquid protein carrier and splashed on leaves of fruit and nearby trees has been used extensively in the past as a killing agent in conventional programmes.

For those opposed to such organophosphate poisons the future may hold promise where certain insect growth regulators or very low toxicity poisons (e.g. boric acid or hydramethylnon) can be substituted. Any such non-contained poisoning programs must carefully consider the risk to non-target organisms and the environment generally.

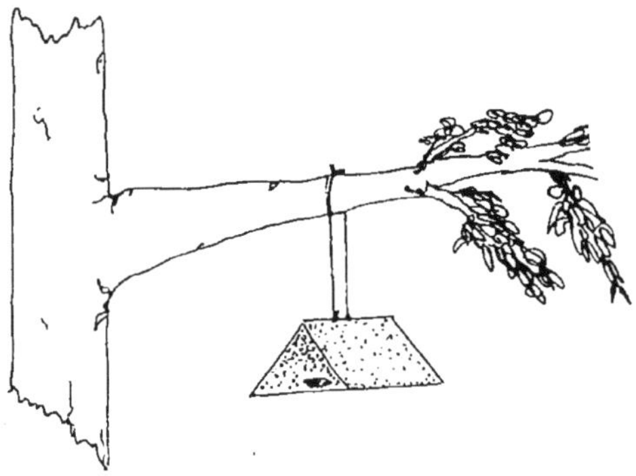

The use of pheromones to attract and trap pests is a useful strategy for the garden.

Repellents

The possibility exists for certain natural low-hazard substances to be applied to fruit at risk in order to deter or repel the female fruitfly.

This is an area where scope exists for careful experimentation. Some formal research has indicated that both olive juice and southernwood tea when sprayed on stone fruit has inhibited egg laying.

Certain other materials (for example, neem, delta-limonene and wormwood tea), known to affect or repel insects, may be useful in such a capacity if applied in adequate concentrations and frequently enough.

A Slater Problem or a Quail Deficiency?

Last year I suffered a terrible plague of marauding slaters throughout my garden and such were their numbers that I was powerless to stop them devouring everything. This year I discovered the delights of quail when I was given a Japanese quail. I started off feeding him with "budgie mix" and leafy greens.

He wasn't overly enthusiastic until I put him out in the "quail tractor" (an upturned washing basket with a couple of bricks on top) in the vegetable garden. He really got excited when I pushed back the mulch to reveal hundreds of slaters and he truly became a most efficient slater/insect vacuum cleaner!

Such is my enthusiasm for Japanese quail that I now have 9 of them on the job, a mixture of adults and juveniles. I keep the juveniles in a large see-through plastic clothes container with a little straw on the bottom and I have recycled an old fridge shelf for the lid so they don't jump out. I fashioned an automatic waterer out of an old plastic lemon juicer suspended to the overhead shelf with a metal clothes hanger and used an old broken bedside lamp (60 watt globe) which I just laid on top of the shelf as a source of heat and light.

However, I have just been made aware that special lamps are required on chicks as the brightness of ordinary light bulbs could lead to problems with blindness.

Quails are omnivorous, requiring animal, vegetable and mineral to be healthy. They eat small seeds (including canola and millet) as well as fresh vegetables (tomato and cucumber rounds, broccoli and sprouts). Mealworms are a favourite too and they will also eat all manner of insects including crickets, flies and even fly larvae (maggots).

I read somewhere that you should chop the heads off mealworms before giving them to your quail as apparently, if eaten live, their tiny jaws will chew into the bird's crop while attempting to escape!

Like all birds, quail need grit to help them with digestion and for their general health. They love to dust themselves, and sand is quite good for this. A lot of people keep their quail in the bottom of the aviary as they are good at cleaning up spilled seeds.

It is difficult to sex the birds under the age of about 6 - 8 weeks and if you want to be sure of what you are getting, go to a breeder. However, the males do have a very distinctive call which distinguishes them from the females.

Another big plus is that Japanese quails are not affected by 40° temperatures like so many other small animals in our systems and they don't destroy vegetable patches with their scratching as chooks can often do.

This is satisfactory so long as the quail are not walking around in bird droppings. Quail feet are also not made for perching, but rather for walking, so the floor should be soft (not wire mesh).

Water should be available for them at the bottom of the cage and I found that a suspended water dispenser is best, as quails will muddy any bowl of water that is left in their enclosure.

Another feature of quails is that they often boink (fly vertically) when excited so you will need to ensure adequate head room whilst your quails are on 'slater duty' in tractors to accommodate this habit.

Japanese quail lay year-round (200-230 eggs per hen per year) and are fully grown and ready to breed at 6 weeks! Unfortunately, they are not good sitters and you will need to use artificial incubation (or perhaps bantams?). Alternatively, the eggs are highly nutritious and are very good hard boiled, scrambled, in omelettes or pickled. Live adult Japanese quail weigh approximately 200 grams and dress out at about 165 grams. We usually open out the bird by cutting right through the breast to the neck and flattening it out with the back of a heavy knife. The birds are then marinated and barbequed.

Handy Hints for Pest Control

An organic spray to discourage white butterflies (Cabbage White Butterflies)

Boil lettuce leaves in water and pour the liquid over the vegetables. If the rain washes this off, it must be repeated. A longer-lasting effect is obtained by spraying it on, using 1 tablespoon of white oil to 1 litre of lettuce liquid. Alternatively, a wormwood tea can also be made up for a cabbage white butterfly spray.

OR

Early morning, before it gets warm, disturb the sleeping places of white butterflies (French lavenders and scented geraniums) and hit the moths with a tennis racquet as they rise lethargically.

OR

Plant a small branch of wormwood alongside each cabbage seedling.

The wormwood generally roots and can be transplanted or discarded when the cabbages are finished.

Slug and snail control

Margarine tubs containing stale beer and buried three quarters of their depth (especially around broccoli and lettuce) make good snail catchers.

If your beer doesn't hang around long enough to go stale, a mixture of honey, bread yeast and water will do. Sugar and water will also work.

Cockroach bait

Mix equal parts of cornflour, icing sugar and plaster of paris. To satisfactorily bait areas such as stoves, refrigerators and free-standing cupboards, sprinkle powder on sheets of newspaper and slide under those fixtures.

Animal repellent for protecting young trees

Beat 5 eggs, mix together with 600 mL water and 150 mL acrylic paint. You can spray on or paint the stems of any trees badly stripped by kangaroos.

Bird scarer

On a recent visit to a commercial organic fruit orchard I noticed no evidence of damage to trees by parrots and no use of nets. The orchardist told me in his opinion nets inhibited growth of trees as well as being time and energy consuming, creating difficulties for harvesting and so on.

Although some small-scale shooting of birds was done, their most successful bird scarer was a line stretched between two very tall bamboo poles, inserted in or firmly attached to posts in the ground. Various "scarers" were attached to the line such as blue plastic bags and shiny metal lids from fruit tins, but the secret of success appears to be in the height of the line above the trees - at least one metre above the tallest tree.

Slater control strategy

Bob has an abundance of slaters in his vegetable garden and a very thick all-over mulch of all sorts of materials, but suffers very little slater damage to plants. He feels that it is because he overhead waters with big sprinklers and his mulch is always damp and therefore more attractive to the slaters. He feels that a trickle irrigated garden would have a lot of its mulch dryish and therefore the slaters would be attracted to the moisture in the plants.

Slater beaters for pot plants

Cut circles out of scrap shadecloth or old flywire and place them into the bottom of pots before filling them. This allows drainage but stops the little critters getting in. If they can't get in they won't be able to chew away at all of the humus and roots.

OR re-pot the plants, remove or shake out the slaters and jump on them. This is particularly satisfying if slaters have just cleaned up your row of beans.

Slater trap

I have a problem with slaters. They just love mulch and seedlings and mature tomatoes 75-100 cm above ground level. I have found a good way of slowing them down. This is a slater trap which is fairly effective but only in a radius of up to one metre. Best results occur when the trap can be set up in the slaters' natural congregating point.

The trap is made of a large necked plastic container 10-20 cm deep. This container is buried to slightly below ground level after having made a small hole in the bottom to drain water. A piece of wire or stick is threaded through a slice of tomato, suspending it over the open container.

Then, either mulch that the slaters like, or an upturned flower pot is placed over the container. The slaters are drawn to the tomato and fall into the pot and cannot get out.

If you have access to a beekeeper, you may be able to obtain some of the residue after extraction of honey and wax, which is very attractive to slaters and makes an excellent bait.

Fly/fruitfly trap

Use empty 2 litre plastic cool (soft) drink container. You can invert the base up into body of container by placing the base in hot water, but this is not necessary.

The next step is to drill or burn a hole large enough for flies or fruitfly to enter. Yellow insulation tape applied around the rim is an added attraction. Now add bait through neck of bottle and replace cap.

For fruitfly control

Claret works well in the fruitfly traps, with tansy planted under the fruit trees and wormwood nearby.

Organic fruitfly control

General hygiene

1. Pick all fruit as it ripens, do not leave on the tree.

2. Collect all fallen fruit each day. Keep ground beneath trees clear of long grass to facilitate this operation.

3. Don't bury unwanted fruit, or throw it into a pit. Either seal in a plastic bag or soak in water topped with kerosene for three days.

4. Keep trees pruned to manageable proportions and if a tree is not wanted, have it removed altogether.

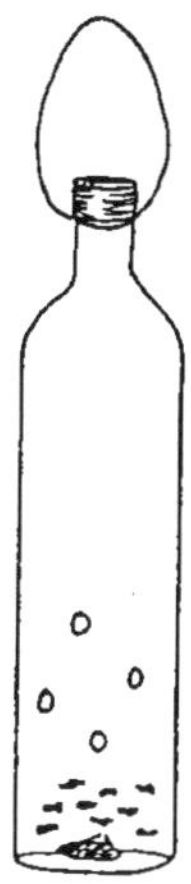

Bait for house flies: kangaroo meat.

Bait for fruitfly: urine and water, claret or other wines, or vegemite and water.

Flies fly up and won't come down, and so drown in water. They cannot become resistant as they do to chemical baits. To empty trap, flush out with a fast hose.

Plastic icecream containers make good fruitfly traps.

Cut a hole on each side a little over halfway up and about 25 mm diameter, and about four or five small holes (6 mm diameter) in the lid.

Fill the trap with whatever lure you use, put the lid on, and hang the container in a string harness in the tree and away from direct sunlight.

Empty traps and recharge them about once a week.

Some observations on fruitfly, from members:

A keen pawpaw grower has noticed a gradual lessening of fruitfly in his garden over the years. He feels it is because when they sting the pawpaws the maggots don't develop, and so it creates almost a sterilising effect on the fly population.

My guavas - normally notorious for fruitfly - are fruitfly free. The 6 trees are all under or near a large Cape Lilac tree, and as this tree is supposed to have insecticidal and fungicidal effects I can only presume that it is protecting my guavas.

An Owner Built Passive Solar House

The basic philosophy behind the passive solar house was to build a low maintenance, low to moderate cost home with minimum running costs, in line with our aim towards self-sufficiency via permaculture.

The design is conventional - perhaps a bit plain - with solar principles applied. It is rectangular with a width:length ratio of 1:2 (not including the garage/workshop) for solar efficiency and simplicity of construction.

Design considerations

1. Orientation. The house has the long side facing north. Probably up to 20 degrees either side of due north would have had little detrimental effect on performance.

2. Windows. Windows are 20% of the floor area, with minimum window area on the south, none on the east, one small window on the west (a concession to views and supervision of kids in the creek), which must be adequately shaded in summer, and maximum on the north wall.

3. Insulation. 50 mm cavity is insulated with 25 mm polystyrene foam. The inner leaf was constructed first and the foam sheets pushed over the brick ties and glued to the inner leaf. The outer leaf was then constructed. The ceiling has not been insulated, but the roof is insulated with a membrane.

4. Thermal Mass. Thermal mass helps to stabilise the temperature of the house throughout the year. We chose to maximise the mass by choosing a slab floor, and common bricks and cement render for internal walls, instead of fastwall or gyprock.

We also included the thermal wall in the centre of the living area. In winter, this absorbs heat from the sun shining on it through the skylight windows, and re-radiates the stored heat during the evening. In summer, the mass of the wall helps stabilise the inside of the house to a cooler temperature.

5. Shading. The eaves have been given sufficient size as to fully shade the large windows during summer. As winter progresses, more and more sun can shine in. The north wall will have a full-length pergola, to support grape vines, which are deciduous. They provide extra shade in the summer but allow full advantage to be taken of the sun in winter. It is worth noting that north wall verandahs are of no use at all in the winter as far as energy saving goes.

6. Curtains and pelmets. Curtains are full length, bunched and thick to hold as much air as possible, as air is the best insulating medium.

All curtains will be pelmetted to stop thermo-cycling, which effectively pumps unwanted heat in, during summer and wanted heat out, during winter.

7. Sealing. Any gaps around doors and windows will allow heat to waste away during winter, and let heat seep in during summer. Particular attention will be paid to external door seals - tape has been used as a temporary measure on the south door.

8. Ventilation. Each room is designed to take maximum advantage of the night breezes and with a larger opening on the north side of the house we have very good ventilation. The skylight windows can be opened to allow any heat build up in the living areas to quickly dissipate. For still evenings, low power ceiling fans are installed in each bedroom.

9. Miscellaneous points. A solar hot water system is a gravity feed unit, with the tank mounted in the roof. It is boosted by the slow combustion stove and has an electrical booster for the occasional quick boost.

Slow combustion stove - hot water boost, heats the living and kitchen areas and is used for cooking. Probably not worth buying a new one, but worth paying to recondition an old one.

Lighting - all fluorescent for efficiency. Disadvantage is in the starting time, which can be overcome with instant starters.

Running a solar home

To gain maximum benefit from a solar house, it must be managed. In summer, all unwanted heat in the form of direct sunlight and radiated heat must be excluded.

As soon as the outside temperature becomes greater than in the house, all curtains should be drawn and doors closed. Only when the temperature outside the house has dropped to below that inside the house, should the house be opened again. Remember that, depending on where you live, many breezes are hot. In winter, curtains are opened to allow in as much sunlight as possible.

As soon as the sun has moved off a window, the curtain should be drawn to conserve as much heat as possible.

Retrofitting House for Solar Energy and Energy Efficient Systems

Retrofit seems to have caught on as the word to describe installing new devices and systems in existing systems. In our case, we are talking about installing a solar energy or energy-conserving device in an existing house or structurally changing the house to make it more energy efficient. Energy efficient housing stems from the idea that if we design our new systems correctly, the solar aspects will be integrated and harmonious and, moreover, less expensive than conventional (fossil fuel) systems. Conventional systems are invariably “retrofit” in appearance - being a “bolted on” device such as a gas heater and air-conditioner and rarely look harmonious with the design of the building. With retrofit solar energy systems, there are many possibilities, some of them simpler or more aesthetically pleasing than others.

Here we will look at the ones considered to be "economical" in some generally accepted sense of the word.

Firstly, the nature of existing buildings must be considered. There is a wide variety of types and styles of dwellings in any region. Some of them lend themselves easily to retrofitting; others are more difficult, as they impose such severe constraints that the owner may have to live with it, or shift.

Many houses are "double brick" and tile - that is the walls have both an inner and outer leaf of bricks with a 50 or 75 mm cavity between them and a roof of tiles made from some form of masonry - generally clay or cement.

There are, however, many houses of other types, ranging from brick veneer to rammed earth for walls, and galvanised iron or colorbond sheeting for roofing. Each of these types presents some advantages for retrofitting, and some severe disadvantages.

The change in building techniques during this century has brought different styles into vogue and different problems for retrofitting. Older houses have ceilings which are 3 to 3.6 m high which not only have a good cooling effect in summer, but also large walls that can be used to good effect. They are also harder to heat in the winter.

The older houses also have wooden floors raised off the earth. Although there are some thermal advantages of the modern concrete raft floor, the wooden floors have a pleasing appearance, which should not be overlooked.

Roofing methods have changed and open eaves are rarely used today. In many houses there are no eaves, while boxed eaves - that is, with a sealed overhang - undoubtedly reduce access to the birds, cats and vermin, but also reduce access for retrofitting, as we shall see later.

Each of the building materials and styles present a different challenge. However, the variety is not infinite and a limited number of techniques could cover most buildings.

Secondly we look at the problems of existing buildings. Aside from the problems presented by the materials and style of house, there are problems that can be common to all houses:

Age

Old buildings can be lovely, but they can present considerable difficulties when the owner tries to make them comfortable. Quite often the materials are fragile - at least to the extent that further load-bearing or cladding cannot proceed without major repairs. Non-standardising of sizes can be annoying. The size of timbers used in old buildings was often whatever was available and may vary from house to house and from place to place within a house. Distances between beams vary, as do wall cavity widths. Extensions and alterations quite often block easy access to parts of the roof structure, or may block the access of sunlight to the room where it is most needed.

Solar access

This is a most vexing problem. A wall or a roof that is otherwise perfect for a solar device may be rendered useless by too much shading. This must be borne in mind when buying a house or planting a tree at any point north of the wall or roof. By using a sun-angle chart, trees can be pruned to suit your purpose. Of course, the ideal situation is to have deciduous trees to the north of the house.

Your neighbour's house and trees may be your biggest problem. As for trees, you may be able to negotiate with your neighbours and convert them into sun-lovers as well. If your neighbour's house shades your wall, which is the most common situation, you will just have to use your roof for solar retrofitting.

If the lot next to you is vacant, beware! A new building may ruin the effectiveness of your solar heaters. Contact your local council and demand to see the plans of the proposed building and make as big a noise as possible. It may not help you, but it will get the council thinking so that others may be helped in the future.

Although high fences are said to make good neighbours, they can also make for poor solar performance. Again, consult with your neighbour and think carefully before constructing new fences that are close to the north side of your house.

Available space

Walls which heat up or transmit heat (solar walls) are often a good idea because they do not make any demands on ground space. A greenhouse can, however, present problems simply because there is no available space to site it.

Many shires have set-back requirements that mean one cannot build within a certain distance of the boundary. Many houses are built out to this point, which means that further building in that direction is prohibited. Again, consult with your council before going too far.

Building restrictions in the other directions are generally the prerogative of the owner and decisions have to be made to uproot flowerbeds, patios, swimming pools and the like.

Suitable site on the house

Even if the above problems have been solved, the house itself may present problems. Solar walls can be placed around windows to great effect, with a little extra work. However, all service ducting should be avoided.

Water pipes, drains, telephone conduits, gas pipes and electrical fittings should all be avoided. Aside from the official attitudes towards obscuring servicing of utilities, it may be downright dangerous to have a gas pipe or electric cable under a solar wall with temperatures rising to 60 - 70 degrees on dark surfaces. There is no reason why a Lawrence or Trombe wall should be in one slab on a wall. In fact, forming panels to avoid pipes may improve its appearance by breaking up the large, blank area of the absorber.

Orientation of house and roof

Walls can be used quite successfully if the orientation is less than 45 degrees from North (the performance will drop to about 70% of that for true North at 45 degrees from north).

Naturally, if a roof-mounted device is to be used, it will work best if there is a part of the roof that faces north at more than 45 degrees. Some houses seem almost impossible to retrofit with panels being mounted against the slope of the roof, which would seem almost prohibitive if 10 to 20 square meters of panels are required. In those cases, a log fire may be the best form of solar energy to use!

Steps Toward Energy Efficient Retrofitting

Insulation

Even a well-insulated house has a fairly large heat flow - requiring in the order of 20 square metres of solar panels (eg. northern windows, Lawrence wall, air heating panels and so on) for winter heating and perhaps some form of extractor fan for summer cooling.

An untreated house will have heat gains or loss at least several times as great, usually making it impossible to find sufficient wall or roof to install solar devices upon.

The value of insulating and sealing a house very much depends on the assumptions one makes. Very few people in this country would try to completely heat or cool a totally uninsulated house. However, compared with that extreme, insulating the whole house would pay for itself within a few years and insulating the ceiling - which in wintertime is really only a thin sheet of plaster with an air film - is the best value of the lot.

Sealing

Air infiltration is a surprisingly significant proportion of the total heat transfer. Even at the very low figure of one air change per hour, (1 a.c.h.), it represents about one third of the total. The average house in this country has not been properly measured, but estimates put the figure at up to 4 a.c.h.

At that rate, the infiltration losses could be approximately 90 kWh/day - about $20 worth of electric heating per day.

Proper sealing of doors, windows, old fireplaces and plumbing would pay for itself very quickly. A full range of weather sealants are available at hardware stores and should be considered as the next project after roof insulation.

Shading for summer

Shading is very important for summer. Sun angle charts should be used to determine the placing of shades and shutters. Broadly speaking, all windows should be shaded from direct gain after mid-October until mid-March.

This requirement is not symmetrical around the summer solstice, so some compromise - generally on the side of more shade - should be reached.

The use of trees is tricky, but worth investigating. Trees not only provide cooling by shading, but also by evaporation, and estimates of several degrees in temperature from evaporation are considered possible.

Deciduous trees and vines are good value. They are generally more shady in summer than evergreens (like eucalyptus) and have a further interesting feature, in that they tend to hold their leaves until the cold sets in, and are a creative way of overcoming the asymmetry in the seasons mentioned above. Conversely, they tend not to come into full leaf until later if winter drags on. Trees are a very good retrofit strategy!

The colour of the building is very important. A dark roof may be 20 degrees hotter than a light coloured roof. Without roof insulation, this may be very significant.

Even with insulation, a dark coloured roof will require about 50 mm more insulation than a light colour to obtain the same heat transfer. In some old houses it is almost impossible to insulate under the skillion roofs with low ceilings.

Painting the roof white or light coloured can have a dramatic effect. This can be important for walls as well, particularly on the west side, or even more so if the wall cannot be insulated. Care should be taken when painting surfaces. The glare from sloped roofs and walls can be most annoying, particularly to neighbours.

Turning our attention now to types of retrofit solar systems, we see that there are numerous ways in which a house can be retrofitted for winter heating. Here are some of the better known and tried types.

Windows

The addition of more windows to the north wall of the house is often overlooked as a means of heating the house. Windows are so common that one often forgets that one of their functions is to let in heat and light as well as to ventilate and provide a view.

A window is much more efficient than any other collector - about 80% or so compared to 30% - 40% for indirect systems.

However, the losses are also greater, and should be taken into account. These losses are not only from the increased conductivity compared with wall material, but also from increased infiltration, unless the windows are permanently sealed.

Solar walls

Converting a sunny, north-facing wall into a solar wall is an attractive idea. In practice, it is not expensive in materials, and the time taken will depend on the structure of the house. There are two basic types:

a. Trombe Walls

The Trombe Wall is a solid (up to 400 mm thick) construction and can either be vented or unvented. One can choose to demolish an existing wall and replace it with a solid wall. This would be a desirable strategy for a frame house that has a low thermal mass. In a cavity wall house, one could simply fill the cavity and clad the outside.

This is a fairly drastic step, which should only be undertaken after consultation with the council and a builder who knows something about water penetration of walls. For vented Trombe walls, the gap should be 1/20 the height of the wall - about 100 mm in a modern wall (2.4 m) and vents about 1% of the wall area.

If the gap and vents are too small, unnecessary heat losses can occur due to re-radiation from over-heating. If the vents are too small, the benefit of a vented wall ie. daytime heating, will be reduced.

If one has a solid stone north-facing wall, it would be ideal for a conversion to a Trombe wall.

b. Lawrence Wall

The operation of a Lawrence Wall is simple. Retrofitting these walls is very attractive because many houses are double brick. Some difficulties can be encountered in practice. Cladding the wall with clear fibreglass reinforced plasterboard (FRP) is advised as it is cheap, light and easy to handle and trim.

Modern FRP is generally impregnated with ultra-violet inhibitors, which vary in effectiveness and are generally not as good as the manufacturers specify - particularly at elevated temperatures. One can choose between corrugated or flat FRP.

The corrugated type can be applied directly to the wall with neoprene washer screws and rawplugs. It has the advantage of not buckling when it expands, but may not look attractive in some settings. The flat type can look quite unobtrusive, but requires 20 mm x 20 mm wooden strips to be bolted to the wall first.

The major problem with retrofit Lawrence Walls is sealing the top and sides of the cavity. Doors or windows will serve nicely to block the ends. Otherwise ingenuity is required.

Flame-retardant polystyrene foam can be used, cut to the width of the cavity, and pushed as far down the side as possible. Several places will form a "labyrinth" seal. The top of the gap can be filled with polystyrene and builders plaster poured over it to form a good seal.

The better the seal, the better the wall will work, particularly in breezy weather when the delicate thermosiphon effect on the wall can be disturbed.

Tiled roofed houses are much easier to retrofit than those with galvanised iron, as it is generally easier to remove a row of tiles along the cavity than it is to crawl into the roof space, or remove the galvanised iron for outside access.

The vents inside should be as large as possible - even larger than in a new house Lawrence Wall - to compensate for any gaps left elsewhere. If the wall already has vents, they should be enlarged on the inside, and outside vents should be sealed.

Even if the wall is not perfectly sealed, it will work reasonably well, as the outer leaf of bricks will get hot and radiate heat across the inner leaf - a phenomenon already well known on most unshaded walls!

Attached greenhouses

A. As a Lawrence or Trombe wall, by painting the wall black and sealing it as for a conventional Lawrence wall. The thermal mass of the wall is therefore utilised.

B. As an air heater. This requires either an open window or upper and lower vents into the house. As the greenhouse provides a pleasant scene, a door or picture window is most effective. Fans can be used to supplement the natural airflow.

There has been much apprehension about the summer performance of attached greenhouses, but there is no problem if the greenhouse is properly vented - by generally having an insect screen door and large upper and lower vents to the outside. The very wary can use shade cloth, deciduous vines or trees or perennial or annual plants, such as sunflowers, to assist in summer shading.

Construction of these greenhouses is not difficult and a variety of materials, such as clear plastic sheeting, can be used. Many free-standing, aluminium-frame, glass or plastic greenhouses are available commercially.

The greenhouse is more than just a solar collector. It can be used as a herbarium or vegetable growing space, as a social space and as a pleasant interface between the inside and outside world.

Air heating panels

These can be attached to walls, under windows or on the roof. As they have no thermal storage in themselves, they can be used for heating the house during the day, or rock storage can be added, according to the ingenuity of the owner. They have the advantage of being able to be placed on the roof in situations where wall and ground space is not available. Fairly simple ducting and a small fan from the roof to the ceiling is required.

Quite a lot of development has gone into window or wall mounted thermosiphon types and a variation is shown below.

Attic space heaters

A method that has been tried in the USA (but not that well documented) is the attic space heater.

People who live in uninsulated houses are familiar with the heat that can be conducted through the ceiling from the air that has been heated in the space between the ceiling and the roof. This heat gain is generally undesirable and eliminated by ceiling insulation.

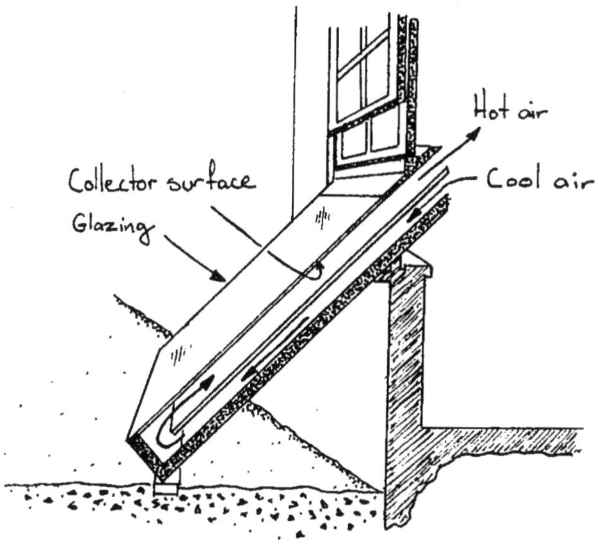

This heat gain can be controlled and even enhanced. The general idea is to paint the north-facing roof as dark a colour as desirable ("mission brown" will do) and isolate that section of the attic space with plastic sheeting, plywood or rigid insulating board.

The ceiling then requires about 100 mm of insulation to offset the extra heat gain in the summer. An extractor fan (a standard ceiling type will do although a slightly more expensive bi-directional fan could also be used for summer cooling) can then be used to pump hot air into the house on sunny days in winter.

Metal roofs are more effective with less leaks, but tiled roofs will also work.

It is not a particularly efficient heater, as it is unglazed and leaky, and there is no direct storage. Indirect storage is only about 25% as effective as thermal mass placed directly in the sun. However, it is cheap and will provide some warmth.

Mudbricks or Rammed Earth?

The question is often asked, which is better: mudbricks or rammed earth? There are advantages and disadvantages in each and it might be sufficient to say that one building material is not better than the other - they are just different. There are three main types of earth walls: mudbricks, pressed blocks and rammed earth. This article only discusses the first two.

Mudbricks

The term "mudbricks" conjures up images of people stamping around a large mound of clay and adding straw and water to make the required consistency. This is a technique which has been used for literally thousands of years and it is still an effective way of building a house cheaply. Furthermore, the rough surfaces and some edges with straw sticking out adds a certain appeal and charm to the house design.

Adobe, or mudbrick, is a very popular earth building method. It doesn't require a lot of technical knowledge, the bricks are generally faster and easier to lay than conventional bricks, and the method does not usually require screening of the raw material. On the other hand you have to wait until they are dry before you can handle and lay them.

Material for adobe bricks is placed into moulds, pushed into the corners and then levelled off with a straight edge (or your foot). The mould is then removed and placed nearby ready for the next brick. Every now and again, about every fourth or fifth brick, you have to wet the mould so that the mould comes away easily as it is lifted upwards.

One person can make over one hundred bricks in one day, but it is exhausting work. Large bricks, with a size such as 450 x 300 x 100 mm can be laid as one thick wall.

Pressed soil blocks

Mudbricks can now be made by using a machine such as a hand-operated cinva ram which produces a mudbrick under pressure. The use of machines will increase the rate of mudbrick production and save you considerable energy. Bricks made this way are much smaller, as a human cannot supply enough force to compress a large brick. You would use these in a typical cavity wall situation.

Machine-made bricks belong to the third type of earth building materials. This can be described as a combination of the two other methods - what is produced are earth rammed bricks or pressed soil blocks.

Here, soil is screened, then cement and water are added and the resulting mix is compressed by a hydraulic ram which pushes the clay and gravel fines mixture into a very compact, weather resistant and stabilised brick which often measures about 350 x 250 x 125 mm. Larger machines are needed to produce bricks of this size.

Bricks made in this way are less likely to crack, have exceptional thermal qualities to insulate your home and are similar sized, which makes the laying of them accurate and easy. Although most Shires require that some sort of weather sealing is used, these bricks do not usually need a protective coat. Your mudbrick house will not wash away in a bad storm or minor flooding. You do not even have to have the customary wide verandas usually associated with earth walls. You may want to have verandas for other reasons though, such as protection from cold wind and rain or from the damaging rays of the sun.

Some mudbricks can be load-bearing and they will need to be tested to satisfy the relevant local authority that they can be used to hold the weight of your roof. Machine-made bricks are stronger than hand-made ones and are more easily accepted as load-bearing.

If you build with a cavity on the outside walls you can pass water and gas pipes, and electrical cable through it relatively easily. When building a rammed earth or pisé wall you have to place the cable, pipe or conduit as you go. Even though rammed earth walls are quick to erect, formwork and retaining bolts and brackets have to be made or purchased.

The choice then becomes, what do you prefer - what do you like? Each earth building technique has its own unique advantages and each meets the needs of the owner and/or builder in many different ways.

Building with Rammed Earth

Growing numbers of environmentally aware consumers are seeking earth-friendly solutions to all aspects of living and lifestyle. This world-wide movement is, among other things, being reflected in the natural tones and textures now fashionable in home building and decorating and the increasing popularity of rammed earth.

Modern rammed earth construction is a desirable, environmentally sensitive housing solution. While its history is ancient, today's technological advancements have strengthened its quality and enhanced its performance.

Mechanisation of the ramming technique has now made rammed earth an economically accessible option to today's home builder.

Rammed earth is a one-step building process whereby prepared earth is placed into formwork and rammed on site. The formwork is then removed to yield the finished wall in its final position.

Fired clay bricks on the other hand (and mud bricks to a lesser extent), require many steps to make the brick and then to lay them.

The use of the world's fossil fuel supplies to produce fired clay bricks amounts to a gross waste of energy. The energy required on average to fire one clay brick in Australia is enough to propel a 4 cylinder car five kilometres.

The 2.1 billion bricks manufactured in Australia in any one year emit at least 1.5 million tonnes of carbon dioxide into the atmosphere. Rammed earth uses l/700th of the energy in the ramming process compared to that required for firing clay bricks.

Unlike clay bricks, earth buildings are "sun dried" and can be made stabilised or unstabilised. A stabiliser is an additive, typically 5% cement, which improves the strength of the walls. This cement content is extremely low compared to other cement-containing building products.

A further energy saving can be made in many cases by using earth from the building site rather than having to transport earth from elsewhere.

Using on-site materials not only provides a demonstration of the local geology but adds to the sense of "belonging" in its location. This would also provide some cost benefit although materials are the lowest cost item in rammed earth building with labour providing most of the cost, followed by equipment and materials. With standard clay brick construction, labour and materials are roughly equal in cost. The low cost of materials would seem to make rammed earth an ideal owner-builder option. However, producing a professional finish can be difficult for the inexperienced.

Source materials

Materials for rammed earth walls are commonly taken from on the site where the building is to be constructed. Since only a narrow band of soil under the topsoil is suitable, it is not necessary to dig a deep hole in the ground. The shallow digging is then backfilled with the topsoil and after one season of regrowth, the site is often indistinguishable from the surrounding ground.

Trees remain undisturbed and such is the harmonious nature of rammed earth that walls look as if they have "grown" out of the ground. There is no other single building material which blends so perfectly with the natural environment. Most often, rammed earth walls reflect the colour of the surrounding landscape.

Natural choice

Rammed earth walls are built as articulated free-standing masonry panels. This method of construction is eminently suited to earthquake-prone areas or problem sites where ground movement occurs due to clay, sub-soil movement or other foundation problems causing subsidence.

Where the panels are articulated, movements of the ground are better accommodated than in conventional brick buildings.

In cyclonic areas, rammed earth buildings are engineered to withstand extreme wind conditions and many successful rammed earth buildings attest to the material's suitability and strength in these designated areas.

Durability

Well-constructed, rammed earth walls can last for centuries. In many parts of the world, including China, Europe, North Africa and Central and South America, rammed earth buildings continue to give excellent service after many generations of occupancy.

Rammed earth will not wash away or disintegrate with time any faster than other masonry systems. The walls are inherently waterproof because of their high clay content and dense compaction. Surface coatings can be applied to further waterproof the walls, but this is not essential unless the walls are very exposed to weather.

The construction process is environmentally-friendly with low energy consumption both during manufacture and through the life of the building. The finished walls are as natural a product as can be imagined with a highly tactile and comfortable appeal.

Thermal properties

As people are becoming more aware of the good feeling of living in a rammed earth home and taking note of the environmental benefits, so the popularity of rammed earth is increasing. Among the major benefits of rammed earth are its desirable thermal properties which are the result of the good balance it achieves between resistive insulation (U factor) and capacitive insulation (thermal mass).

The U factor of a building material refers to its ability to prevent heat being conducted from one side of the material to the other. Thermal mass refers to the amount of heat a material is able to absorb and thus the amount of heat required to raise the temperature of that material.

Rammed earth has quite a high thermal mass and is thus able to absorb a lot of heat without significantly rising in temperature. The high thermal mass of rammed earth contributes to a comfortable environment by evening out temperature fluctuations.

Acoustic properties

Rammed earth, due to its density, is an excellent barrier to airborne noise transmission. The textural face of rammed earth walls gives excellent reverberation characteristics giving a pleasant living environment.

Humidity equalisation benefits

Research has shown that rammed earth walls can moderate humidity changes by storing and releasing small amounts of water vapour in the home thereby increasing year-round comfort.

Other properties of rammed earth walls

Adding to the comfort and good health of rammed earth houses is their non-toxic composition, excellent acoustic properties and the avoidance of any hazardous electromagnetic effects from steel frames (or reinforcing) or insufficiently protected electrical wiring. Steel framed houses act as a Faraday cage, moving through the earth's magnetic field and producing electromagnetic energies unpleasant to humans. Thick rammed earth walls also provide good screening for the weak electromagnetic fields generated by electrical wiring.

Rammed earth walls can be constructed in a wide range of colours and textures from white coastal limestone to deep earthen reds and browns.

Surface texture can also vary from a smooth render-like finish to a highly textured gravelly finish, and everything in-between. Patterns can be indented into the walls or motifs made proud to provide surface relief.

So, with its superior structural, thermal and acoustic properties and natural good looks, rammed earth is proving to be an ideal modern building material. But there is more. Rammed earth is probably the single lowest environment impact building system that is commercially available today.

Organic Solar Architecture

The synthesis of straw-bale construction and renewable energy

Using straw for construction of buildings has been going on for centuries. In many countries around the world, reeds and straw have been used for thatching of roofs and walls; in some places it is even used for the floor.

In the late 1800's the Industrial Revolution saw the invention of a Hay Baling machine, which enabled straw to be formed into compact, easily managed and stored bales. Faced with a shortage of building materials, farmers in the sand plains of Nebraska USA began creating their homes with baled meadow grass, which grew in abundance there. This began what is now called Plastered Straw Bale Construction. Many houses and structures were built in this region between the 1890's and 1940's.

In the 1980's, some 90 years after the first straw-bale buildings, building with straw bales has been revived and has spread to other countries around the world including Canada, Mexico, Latin America, France, Scandinavia, United Kingdom, New Zealand, Mongolia, Russia, and now Australia.

Building with straw bales, when compared to other forms of building construction, offers a considerable net saving. When integrated with renewable energy systems, good organic design and appropriate concepts, buildings and environments can be created which are beautiful, energy producing, low cost, healthy, organic and sustainable.

Construction method

Building with straw bales is essentially a system of constructing walls with bales of straw and rendering (plastering) them over with some protective compound. Simply put, you construct a strip footing the width of the bales, with steel starter bars protruding out of the concrete.

The first course of bales are then impaled over these bars. The remaining courses are stacked in the same way as bricks and tied together with short lengths of steel stuck through the bales. Bamboo can be used in the place of steel and is in fact stronger than steel, weight for weight.

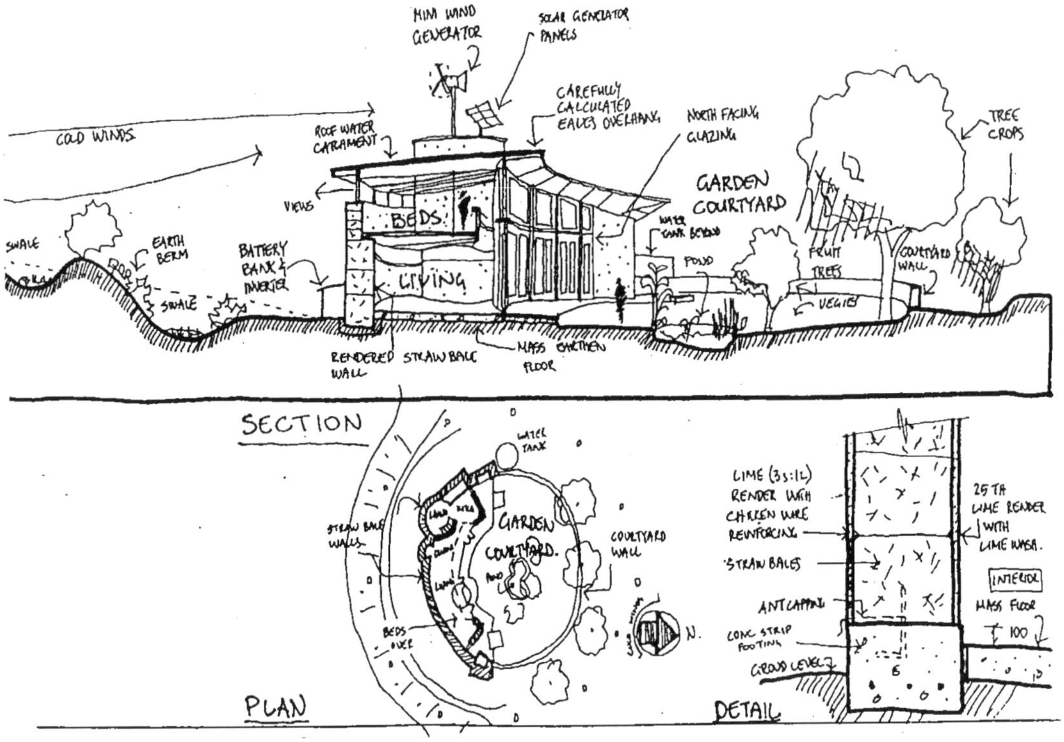

Door and window frames are installed as you progress. When the full height of the wall is reached, a wide top plate is placed on the top of the wall and fixed to the footing via tie down rods or cables. The roof can then be attached to the top plate. For non load-bearing walls, the roof framing would be attached to a post and beam system and the straw bales stacked between the posts.

After the bales are tied down and allowed to settle a bit, a lime plaster, of 2 to 3 cm thickness, is applied to both inner and outer surfaces. This protects the bales from moisture, vermin and fire, and provides the base for the eventual colour finish. Finishing the bales with a lime and sand render is a recipe that goes back thousands of years. It yields a soft, fire-proof, sanitary finish that allows the walls to breathe, creating an interior which has a healthy fresh air feel about it.

We now have a finished wall. This deceptively simple system offers unique benefits to the astute home owner and/or builder.

Straw is an annually grown material. The combination of sunlight, water and soil are the main requirements for growing long stemmed cereal crops. After the grain is removed, the long stems are baled and transported ready for use in a building. The energy required for producing straw is mostly natural, resulting in conservation of scarce earth and energy resources. Vast areas are dedicated to the growing of wheat, barley, rye, rice and other cereals. It makes good sense to use the by-product of this enormous industry for low-cost housing.

Straw is a very forgiving building material, allowing you to cut, move and trim the light-weight bales as required. As no great skill is needed, you can build with the help of your family and friends. This allows money to be saved on labour, resulting in lower construction costs and smaller debts.

Of very high insulation value, straw bales create a stable interior environment which is easily heated and cooled and provides far superior living conditions to that of other common building materials. It must be remembered that straw bales are a compressible material, so keep your roof spans to a maximum of 6 m. This will mean that the building will be kept long and narrow.

The maximum length of an unsupported straw-bale wall is 6 m, so you will need either corners, attached piers (buttresses) or the like if the wall is longer.

Wall height for a 450 mm high bale is a maximum of 8 bales if unsupported. You will find during construction that from the fifth course up the wall will have the wobbles. This is typical and is rectified when the top plate is fixed down.

Next orientate the building for solar penetration. To increase sun penetration, whilst keeping your openings to a minimum, I suggest that a clerestory window be included above the northern wall. This will allow direct sunlight in during winter, providing plenty of warmth and light.

In summer, the eaves overhang and shade the glazing. The clerestory improves and creates controllable natural ventilation, and also allows the much-loved verandah to be included, which traditionally has created cold interiors in winter. This verandah may also be a shadehouse, glasshouse or pergola.

An idea I suggest to many people and one which helps to keep costs down is to create a number of small simple buildings and group them around a central courtyard, linking them all with a verandah.

Courtyards are great for creating different micro-climates, particularly for tropical fruit trees, for increasing ventilation and creating wind protection and privacy.

The low-cost nature of straw-bale walls makes this kind of design reasonably cheap.

These simple design ideas, in combination with a solar and wind energy system, can begin to establish the basis of a highly integrated building energy and food system.

These can be adjusted and modified to suit the individual requirements, the lay of the land, local climatic conditions and so forth.

The effect nevertheless, is of the garden coming into the interior, sun penetration being controlled, healthy ventilation, privacy and security, beautiful interiors connected to the exterior, a thermally efficient building and delightful living spaces all year round.

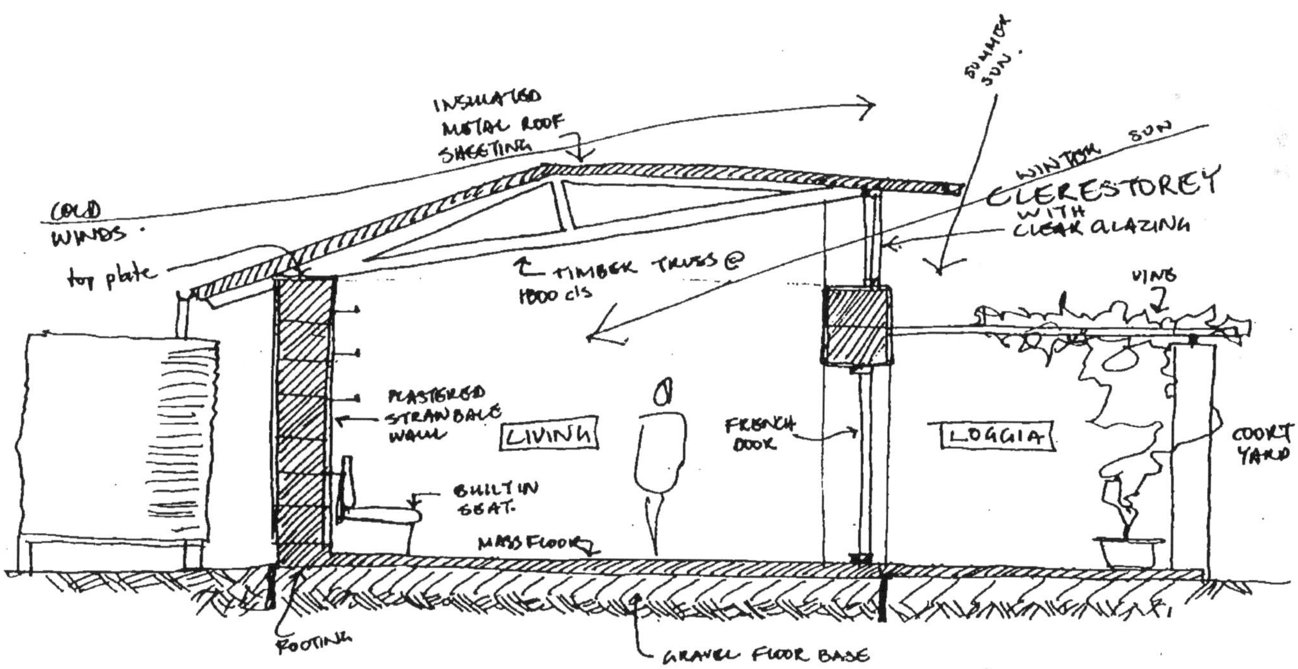

An Integrated Solar House

Solar houses are not new. Many permaculturists have built passive solar homes, but few contain the features of a special house in the hills near Perth. While it has the standard solar hot water system and orientation towards the north, it also includes a range of features such as an active hot water storage tank, solarium with solar pergola and solar air panels. What's more, the house is made from rammed earth bricks.

The passive solar home

The energy needs of a home can be solved by a combination of design and materials. Good design permits passive solar gain, and building materials and hardware allow energy capture, storage and transmission.

A passive solar house is one which can effectively utilise the sun's energy to keep it warm or cool and to moderate the adverse effects of climate on the house. This passive solar house includes:

• long axis east-west with the house twice as long as wide, giving minimum exposure of the western wall to the sun.

• about 40% glass on the sun-facing (north) side.

• minimum windows on the east and west sides, with two windows, located on the eastern side, to harvest and direct cool night time summer breezes into the house.

• polystyrene insulation in the wall cavity on the prevailing wind and rain sides - the south (sun-shaded) side of the house and the west side.

• solarium on the northern side. This is a room which has clear polycarbonate (laserlight) roof sheeting and a solar pergola on top of this to allow winter sunlight into the house.

• integrated roof panel. The panel is a sandwich of polystyrene foam between two sheets of colorbond iron. The white underside is the ceiling and the top green sheet is the external roof surface.

• clerestory window bank completely along the roof line. This also regulates the amount of light that can enter the house.

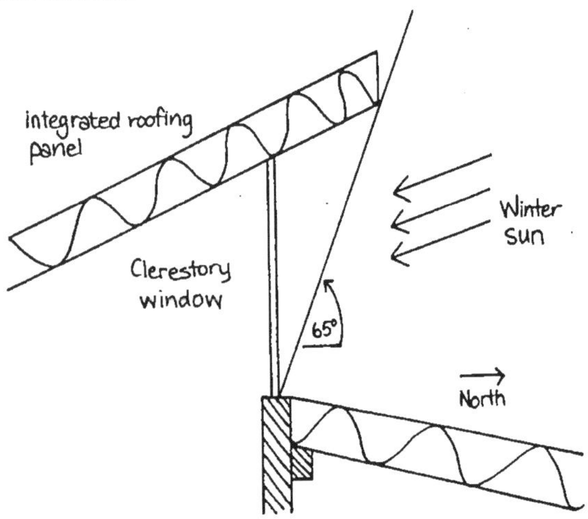

Figure 1. Clerestory windows are positioned right along the northern roof-line.

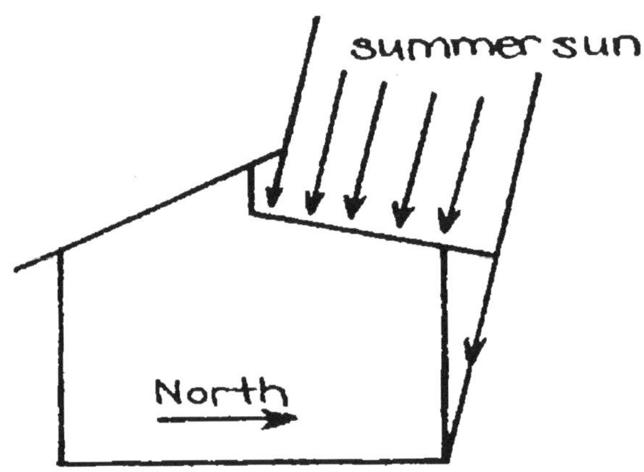

Figure 2. The house is shaded from summer sun by making sure that the roof overhang is sufficient to block direct sunlight.

• bedroom areas on the southern (shaded) side, although the clerestory windows direct winter sunlight into these rooms.

• living areas, including kitchen, study, toilet and bathroom on the northern side. These areas get plenty of warmth in the winter.

• eave overhang calculated to minimise summer sun entry. The roofing panel overhangs the clerestory windows and northern wall so that no summer sun enters the house or even hits the northern wall for most of summer. This principle is illustrated in figure 2.

Solar houses such as this one are remarkably effective. While the outside temperature may fall to 2 or 3°C several times during the colder months and most of winter has temperatures less than 10°C, the house has never dropped below 13°C even during the coldest spells. During winter, the house temperature mainly stays between 15 to 20°C day and night.

Summer is no different. The inside temperature has never risen more than thirty one degrees while the outside cooked at over forty, often for days at a time.

Solar pergola

A solar pergola is an effective way to allow winter sunlight into a house but keep summer sunlight out. Wooden or sheet metal blades are placed at an angle so that summer sunlight cannot pass through, but winter sunlight, at a lower angle, can.

The blades are normally fixed and set, so that maximum light will pass through into a room at the winter solstice, when the sun is at the lowest altitude in the sky. Spacing between the blades regulates when sunlight is able to pass through, so that you could have shading for four or more months during the hottest part of the year. Maximum sunlight penetrates the solarium in the middle of winter. Figure 3 shows this principle in action.

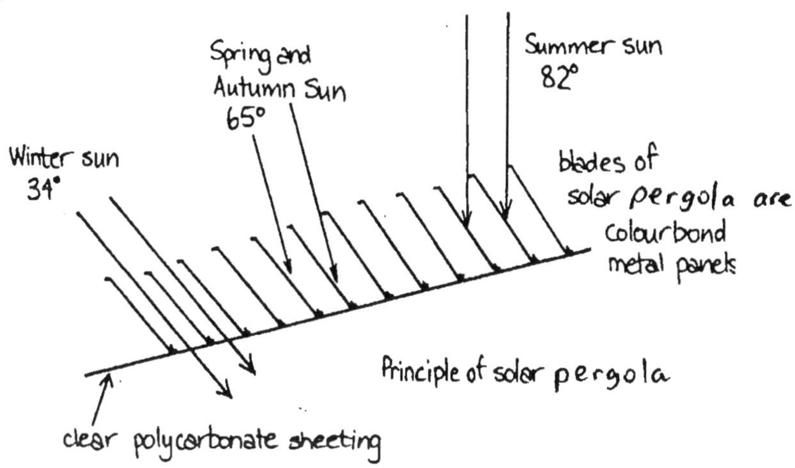

Figure 3. The solar pergola blades only let winter light into the house.

In our climate, the blades are angled at 35° and placed 125 mm apart. This permits complete shading of the solarium from mid October to early March. We chose colorbond blades as they were cheaper, easy to build and lighter than a wooden pergola. The dark slate floor and the rammed earth brick walls, with high thermal mass, absorbs this energy during the day and releases it during the night - into the nearby bedrooms and living areas.

Solar air panels

Solar air panels can be easily made and fixed to a roof. These are similar to solar hot water system panels except that air, and not water, is passed or pumped through them. These can be made and fitted to an existing home as a retrofitting strategy.

During the winter months air is drawn from particular rooms and is heated as it blows across the panel. The hot air is then ducted into the house and can enter two designated rooms. Sensors detect differences between the room and the panel temperatures, causing an air suction fan to turn on or off. The movement of air is thus regulated. Naturally, the solar air system doesn't operate during summer. We have set the controller to switch on the system when the panel temperature is 10°C hotter than the room temperature.

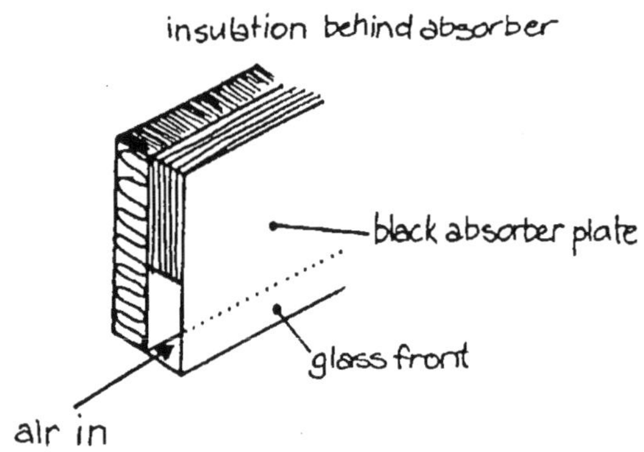

Figure 4. Details of the new technology of solar air heating.

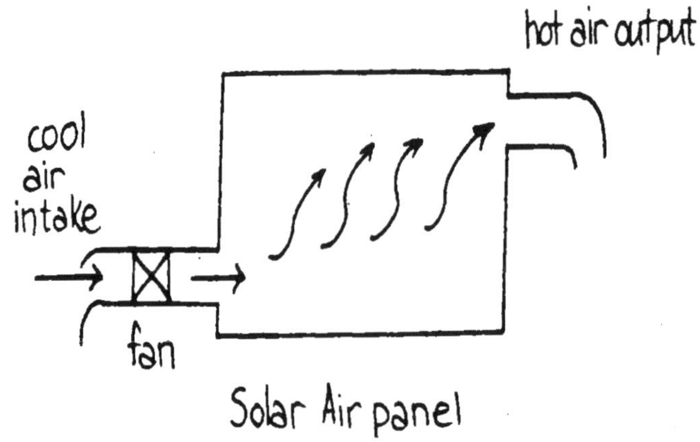

Figure 5. Solar air panels recycle air from rooms in the house.

Active solar hot water system

There are two main types of solar hot water systems. In a passive hot water system, which relies solely on natural convection currents to move water, the storage tank must be slightly higher (300 mm is adequate) than the panel so that cold water from the tank falls towards the bottom of the panel. The collector panel heats the water and the hot water rises towards, and into, the storage tank.

In an active system, energy is used to pump water through the panels to and from the storage tank.

The tank can be positioned anywhere, such as in the roof space above the bathroom or outside on the ground near the kitchen.

Sensors detect the temperature differences between the panel and tank and, when the panel is 10°C hotter than the tank, the pump is engaged and water starts to flow.

The tank is made from two recycled 200 L (44 gal) oil drums. They are welded and bolted together and house a long copper coil at the top of the tank. Our system is a heat exchange, active solar HWS. The water that circulates in the panels and tank remains in the system - it doesn't travel to the kitchen tap!

The principle of a heat exchange system is simple. Cold water (from the mains supply) enters the top of the storage tank and passes through about 60 m of 13 mm copper coil. As it travels through the coil it absorbs heat from the surrounding hot water and emerges from the other side of the storage tank to be taken to the bathroom, kitchen and laundry. A slow combustion stove, with a water jacket, can supply hot water as a booster to the system.

Integrating the house and garden

Ideally, the house and garden should complement each other. The garden can be designed to take an active role in the heating and cooling of the house. For example, deciduous trees should be planted along the sun-facing wall of the house. These regulate microclimate by providing shade in summer and allowing sunlight to filter through during winter.

Part of the permaculture design of the property incorporates gardens with the house. There are a mixture of deciduous and evergreen trees around our house. A sector plan of sun angles indicated where each type of tree had to be planted. Evergreens were planted in areas around the house where only summer sun reaches, such as the south-eastern and south-western walls.

Deciduous trees were planted mainly in the winter sun sector, starting from the western side across the northern side and around to most of the eastern side. These deciduous trees include paulownia, silver birch and a number of various types of maple trees.

These gardens complement the house by sheltering the walls from the hot summer sun and allowing winter light to penetrate the house when required.

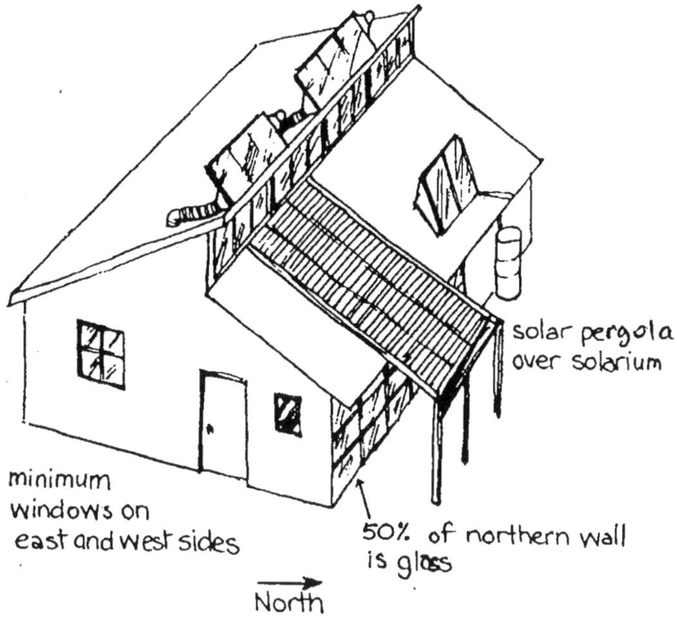

Permaculture is about designing our houses and gardens to be more interactive. It encourages us to find new ways to make our homes more energy efficient and less demanding on limited natural resources.

Careful, thoughtful designs of both our garden areas and houses helps to achieve these aims.

Solar Architecture Revisited

The concept of solar architecture is nothing new. The ancient Greeks utilised the benefits of the sun some 2500 years ago. With the short supply of wood as fuel, they found it necessary to build their houses to take advantage of the sun's rays during the moderately cool winters and to avoid the sun's heat in the summer.

Excavations of many Classical Greek cities show that solar architecture flourished throughout Ancient Greece. Individual homes were oriented towards the southern horizon and entire cities were planned to allow their citizens equal access to the winter sun (northern horizon in the southern hemisphere).

The Greeks venerated the sun and believed that exposure to the sun nurtured good health. Thus solar architecture became a cultural necessity. They built their homes so that the winter sunlight could easily enter the house through a south facing portico similar to a covered porch.

The main rooms in the house were warmed by the rays of the sun streaming in through the portico, but were sheltered from the cold winds from the north.

Today, solar architecture is undergoing a resurgence as more people not only recognise the comfort benefits and lower energy bills of solar architecture, but now see a way of helping reduce the polluting effects of greenhouse gases through less reliance on fossil fuels.

Basic house design principles remain:

- Orientation of the main living areas toward north.
- Glazing used to trap the sun's warmth.
- Thermal mass to store heat from the sun.
- Insulation to reduce heat loss or heat gain.
- Ventilation to capture cooling breezes.

A well-designed solar home should remain within 18° C to 28° C throughout the year and save 60 to 70% in average household heating costs.

Strawbale Building

The advantages of straw as a building material are quite well known by now. Its insulation value is better than anything else, it is cheap to buy, it is a waste product with a new use, it is easy to handle and work with, and creates an aesthetically pleasing ambience.

The first decision to make is whether to go load-bearing or infill. My feeling is that if one is to go to all the effort of building a framework, then there are cheaper ways of infilling the walls. Not only is there the extra expense of the frame, but more labour is involved, more materials are consumed and the job generally takes longer. Pole frame construction with a wattle and daub infill, or part timbered with a solid plaster would be cheaper, less work and you would not lose the beauty of the poles. Strawbale works very effectively as a load-bearing structure, so to complicate it is to waste time, effort, money and materials.

Regardless of the building technique used, the most expensive components of construction are the roof, the footings and fitting out. Walls are, in fact, the cheapest part of a house. What it all boils down to is skilled labour costs. The faster something is to put up, the cheaper it is.

There are certain areas where material costs have an impact, but more of that later. This is where strawbales come in, but like all things that seem too good to be true, there is a catch. To be effective, strawbale must be owner built.

Strawbale is a labour intensive system. There is the laying of the bales, pinning, stitching, connection of the top plate and tensioning. Next a scratch coat of render is applied, then one or two coats of render on top of that. Very labour intensive, very time consuming, and if you are paying current rates of labour for it all, very expensive.

The saving grace, and to my mind its greatest advantage over any other building material, is that any seven fingered, three handed, all-thumbs person can not only do it, but do it well and thoroughly enjoy themselves. It is a building technique that compels people to do what should come naturally, build their own home. It is done in such a way that it has come into its own as a fast, simple and efficient form of building. From children to the elderly, everyone is able to contribute. A truly democratic form of construction.

Now, back to the issue of legs. The traditional way of mounting the strawbale house is either on a concrete strip footing or on a raft slab. It is a pretty standard procedure, mainly because of the prevalence of brick and other monolithic types of construction. Yet I fail to see its relevance to strawbale building. This is where material costs, as mentioned earlier, comes to the fore.

Strip footings for strawbale have to be big. Not because of the weight but because the bales themselves are wide and need to be above ground level. This gives a footing that by volume is almost twice that required for brick, yet carries less weight. If you then run that with an insitu slab you have to take into account excavation, sand fill, compaction, a water-proof membrane, reinforcement, metal supports and so on. On top of this, concrete really needs to be laid all in one hit.

It needs skilled people on site, and then much time and care in curing it properly. All this becomes intimidating and expensive and the money is paid up as one rather large lump sum, which is not really an owner-builder friendly way of doing things.

This leaves the option of a suspended floor on stumps. The advantages of this system are many. Stumps can be timber, pre-cast concrete or concrete cast insitu. It can be done in stages as money and time allows and does not call for great skills other than getting things level and plumb. There is minimal site disturbance, especially if it slopes somewhat, and all work can be done by hand. Because the floor is suspended, termite observation is good, access for services is easy and flexible and under-floor ventilation can now be taken advantage of. The choice of flooring material is wide and varied and the bales are attached in a manner surprisingly similar to that involving strip footings. The stump system works particularly well in clay but is just as at home in sand. If there is any settling of stumps, solutions are easy to apply. The system is flexible, figuratively and literally.

One strategy used in the infill method is to stack the bales on edge. This uses less bales and cuts the cost. But does it? It uses 15-20% less bales, but creates a new dilemma. The render mix will not adhere directly to the smooth sides (top and bottom of bale) without reinforcement. So into the equation comes rolls and rolls of chicken wire, and hours of tedious sewing and stitching. Absolutely no gain, and an added complication and cost factor.

Strawbale building is a fair-weather activity, with speed of wall raising and rendering the essence. If all is planned and organised properly, this process should take days, not weeks, thus reducing the risk of untoward weather interrupting work. Once that first coat of render is on, the walls are safe.

Obviously design considerations have to be taken into account. If a wall is to be higher than usual, have more than 50% of its area as openings, or if the building span is greater than 5 to 6 metres, then a frame structure with infill needs to be considered. But it comes back to designing within the limits of the building material.

All building materials have their structural and loading limits. Load-bearing encourages a smaller scale, builder-friendly, low-cost approach to housing.

This is where another factor in the cost of building becomes obvious - a smaller building costs less than a larger one. Not always in a proportional sense, but definitely in a final absolute cost. How often has it happened that someone figured out that they can build their home for half of what it would cost to build normally, and then been tempted to double the size of the house. It is about time we gave more than lip service to the concept of living and building within our means and needs. Keeping buildings simple and to a human-scale reduces their costs.

The structure of a load-bearing building is totally tied together. The roof is attached directly to the footings. This is also an advantage in areas with high wind velocity. Every bale is pinned to six other bales. The system is not monolithic like brick or rammed earth. It can bend, twist, flex and absorb energy, but not break like a masonry wall. It is the old story of the reed and the oak.

A timber structure works in a similar way - it also has an ability to flex and absorb energy. However, because the wall bales in infill are not tensioned, they are not so able to absorb and thus spread over a wider area the energy and the forces acting on the building. It is still a far more satisfactory method of building than a monolithic construction, but as load-bearing has the proven structural integrity, and the timber framing loses out in the costing, it means that load-bearing is still the superior method.

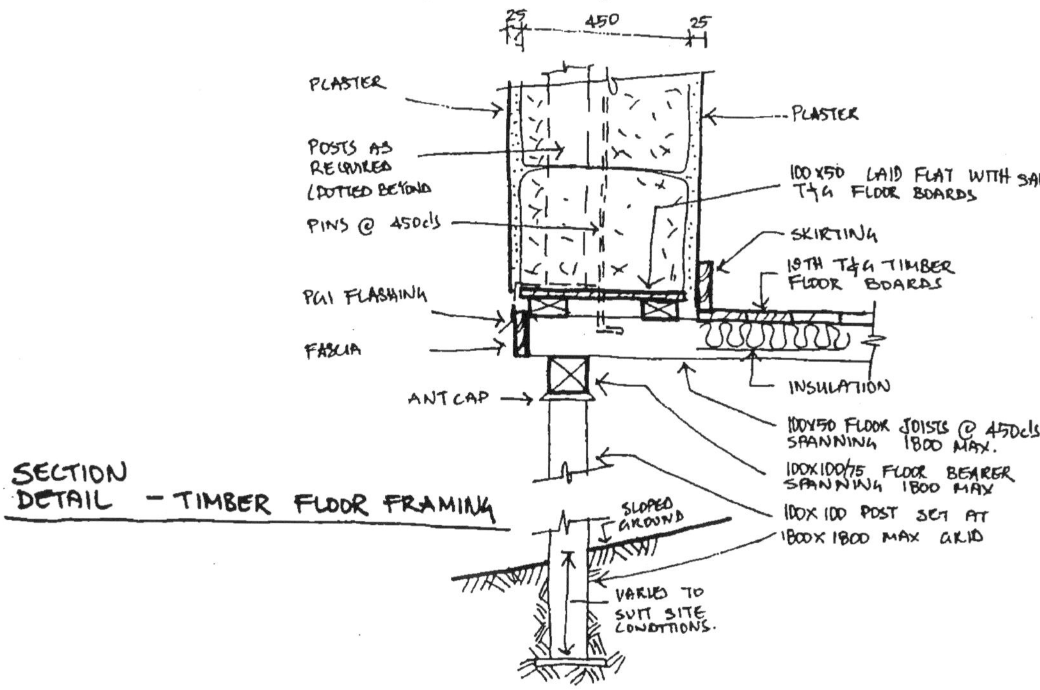

SECTION DETAIL - TIMBER FLOOR FRAMING

What it boils down to is not the building method itself, but the nervousness that afflicts various experts and officials when confronted with something unfamiliar to their training and experience. This is not to be unexpected.

To approve something, they need proof that it does work. The problem for any new or revived building technique is to become officially recognised.

It was like this for mud bricks thirty years ago, and for rammed earth more recently. I guess load-bearing just has to go through a similar gestation period. As someone once said, "The hardest million dollars to make is the first." It is the same for alternative building - the hardest hurdle to clear is the first one. Once you have made it over that first hurdle, then you are in the race to get your approval.

Building Your Own Solar Hot Water System

Permaculture is concerned with designing systems that will run in a world of diminishing energy resources. My aim was to design a solar collector which could be built by myself, with limited equipment and skill, from readily available materials.

Materials

Secondhand 100 litre electric hot water systems, complete with heater and thermostat in working order, are readily available from demolition firms. Other (new) materials needed to make the 2.4 m x 0.9 m collector panel are as follows:

2.4 m x 0.9 m x 10 mm chipboard
Glass, full sheet 2.4 m x 0.9 m
100 mm x 20 mm dressed timber
Glass, strip (option for double glazing), 350 mm wide x 0.9 m long
Solder
Alfoil and fittings
28 gauge copper sheet
3 tubes black butyl mastic
20 mm copper tube x 12 meters
Pipe insulation
Caneite
40 mm x 12 mm wood strips
Silver brazing alloy and flux

Design

A schematic drawing of the complete system is shown. The bottom of the tank must be higher than the top of the panel.

In designing the solar collector, the following facts were considered:

1. Copper tubing is easily and cheaply joined, withstands high pressure and high temperature, and resists corrosion.

2. Glass is ideal for the cover. It transmits 87% of the sun's light (short wavelengths), yet is virtually opaque to the infra-red (long wavelength) radiation which comes from the hot surface of the absorber. Plastics are not opaque to infra-red and hence are not suitable.

3. A flat, black paint absorbs 96% of the light which falls upon it. The energy of the absorbed light initially can only heat the surface.

In operation, some of the heat travels by conduction to the water pipes, some heats the air in the box by contact and some radiates from the surface.

To facilitate conduction of heat to the water pipes, the absorber panel should be of adequate conductivity, the pipes should be sufficiently close together and well bonded to the surface. To minimise heat loss, the air must remain trapped in the box which, therefore, needs to be sealed, except for a small vent at the bottom to allow for expansion. The heat of the trapped air, together with the heat radiating from the black surface, will warm the glass cover which will then lose heat to the atmosphere. The amount of heat lost will be proportional to the temperature difference between the glass and the outside air and the speed of the wind over the surface. These factors suggest that double glazing and the use of selective surfaces may be useful. A selective surface is one which readily absorbs light, but radiates little heat, unlike black paint, which radiates as well as it absorbs.

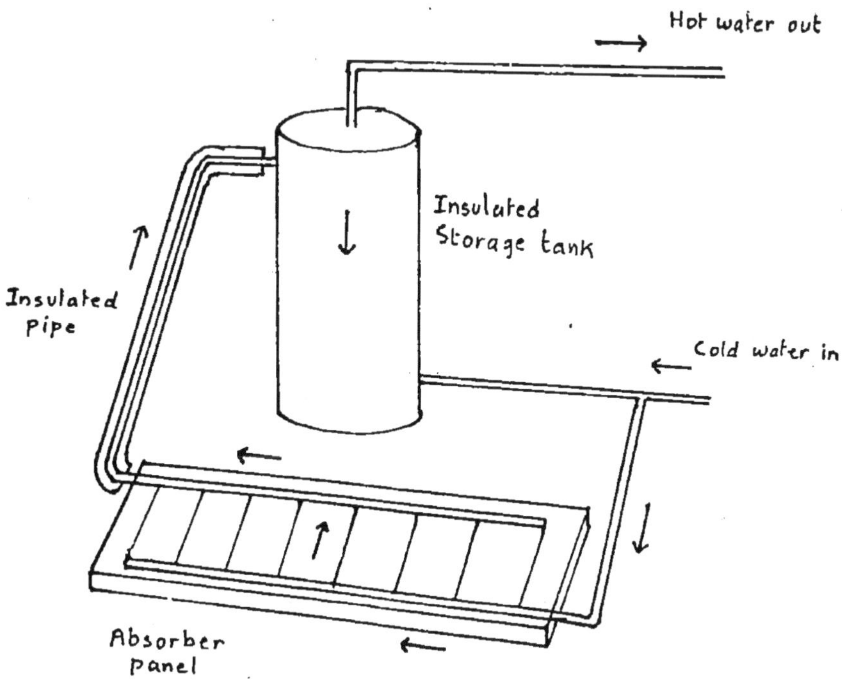

Double glazing is generally not needed but with an air temperature of 20 degrees and water temperature of 60 degrees, double glazing increases the efficiency by about 5% but increases the cost by 20%. The money may be better spent on making the unit bigger. In my own case, having noticed that the upper portion of the glass cover was much hotter than the rest, I decided to use double glass on the top third only.

Construction

The box was constructed with sides of 100 mm x 20 mm dressed timber glued and nailed to a 2.4 m x 0.9 m sheet of 10 mm chipboard.

Strips of timber 40 mm x 12 mm were nailed and glued all around to form a recess for the glass, being extended at the ends to provide handles for carrying the panel. Caneite sheet and aluminium foil were used to insulate the bottom, as in the diagram.

20 mm copper tubing was used for the upper and lower headers, drilled with 10 mm holes at 17 cm intervals, to take the risers of 10 mm copper tube. These were joined with the silver brazing alloy and the appropriate flux, using a borrowed oxyacetylene torch.

The absorber is a 26 gauge copper sheet soldered to the risers with ordinary plumber's solder and painted matt black. It is essential that the hot water pipe from the absorber panel to the water tank be well insulated or the system will not work. All wood surfaces are painted, the inside being finished with aluminium bitumen roofing paint.

Electric booster

If building your own hot water tank with an electric booster, place the element and thermostat about one-third of the way down from the top of the tank.

This will virtually eliminate the need for manually turning the booster on and off as it enables two-thirds of the water to be drawn off before cold water reaches the thermostat.

The Production of Methane Gas

Humans have known about methane gas since Volta discovered it in marsh gas in the eighteenth century. Cheapness of other fuels has taken away the interest, but as the oil runs out it is bound to become popular again.

Anaerobic digestion, the process that produces methane, has at times been hailed as the answer to all our energy problems, and the thought of everyone having their own digester and being self-sufficient in energy often crops up when talking or reading about alternatives to the current energy production systems.

There are three natural sources of methane available and only one is commercially exploited. This is extracted from what we call Natural Gas, from either the N.W. shelf or similar situations. The second is from marsh gas produced by the decomposition of plants by bacteria in the absence of air. The third source is from the stomachs of herbivores like sheep and cattle (i.e. ruminants). It is the animals digestion system and bacteria that break down the cellulose in the grasses and food they eat into molecules the animals can absorb. Hence they belch and bloat when feeding on rich foods.

Sewage plants were once used to produce gas for street lighting and to operate pumps in the early part of the century. Only developing countries continue to use methane digesters.

During the Second World War people in England did build and use digesters successfully. Small-scale operations can work well as they do now in India and China where no funds are available for other fuels. Also the use of the by-products (i.e. the sludge) for fertiliser makes the process a winner for these people.

Some may regard the fertiliser as more valuable than the methane, especially if it is left in the wet state and high in nitrogen. Digestion removes the polluting hazards as well as producing an energy source. Every system and every batch is different in every digester, hence the delicate balance needed in keeping the process going. Problems seem to occur when finding the least energy consuming method, especially when charging and discharging the digester.

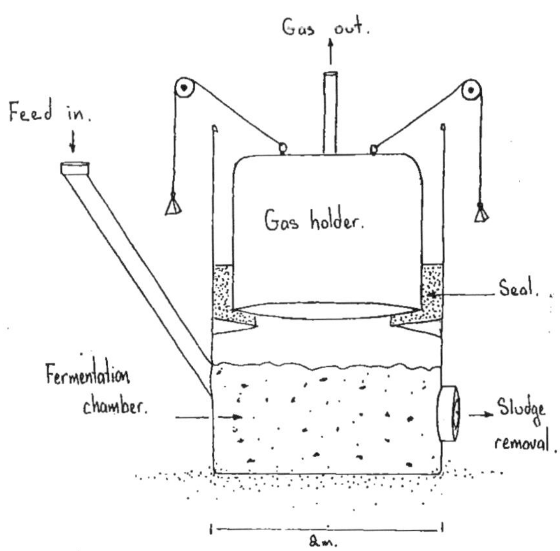

Other inefficiencies and difficulties arise in the strength and volume of waste used. Probably the larger digesters work best because these inconsistencies are more easily balanced out.

The breakdown stages of organic substances are threefold. The first is known as liquefication as the enzymes produced by the bacteria are dissolved in the water surrounding them.

Phase two is the acid-forming stage and the bacteria so far can be both anaerobic (without oxygen) or facultative (with small amounts of oxygen). Acids produced are acetic and fatty acids, like propionic and butyric acid.

Carbon dioxide and hydrogen are also released. By now, most of the anaerobic bacteria's work is complete. Methanogenic bacteria take over from here, the acid-formers having ensured the contents to be free of oxygen.

The progress of a batch can be followed by the acidity changes. Acid-forming bacteria obviously will bring the pH down until the methanogenic bacteria take over and the solution's pH will again rise as the acids are converted into methane. This can take several days or weeks. Unfortunately, ammonia is often produced as well, and kills off the methanogenic bacteria which can only produce efficiently in a pH condition between 7 and 8 (slightly alkaline). Even so, methane production from animal wastes is still the best method of preventing polluted water courses and producing useful fertilisers for the soil.

Simple Food Drying

Drying foodstuffs has been done for a long time. Foods similar to dried fish and meat 'jerky' have been documented by anthropologists in many ancient cultures. Crop drying has been practised for thousands of years and 'sun drying' in open air is still the most common method of food preservation by drying.

For all its popularity direct sun drying does have problems. Foremost is the irregularity of results. This either requires a lot of monitoring (labour) or produces inconsistent results. If one is trying to maximise production, e.g. for a livelihood, it's understandable why other methods, mostly based on hydrocarbon fuels, have become the norm. Many a tomato or apple has been dried just in the sun on an old bedframe or a screen door. Here one strikes another problem in that open exposure allows contamination from dust and vermin or, worse yet, pests actually consuming some of the banquet laid out obviously just for them.

Mechanical dryers were developed in the last century to use cheap and readily available fuels, mostly hydrocarbons. Such systems operate with thermostats which regulate temperature automatically, baffles to automatically adjust air flow and barriers which keep out light and contaminants.

There is no reason solar energy can't be used as the main source of heat with alternative (storable) fuels only used when necessary as back-up; rather like a solar hot water system using a booster device. Given an understanding of system mechanics, solar electricity can even be used to activate solenoid valves and drive motors to move air or baffles. Just such a system was developed for fruit drying in the mid '80s. A diagram of Starcken's dryer which operates on a vacuum principle is shown in Figure 1.

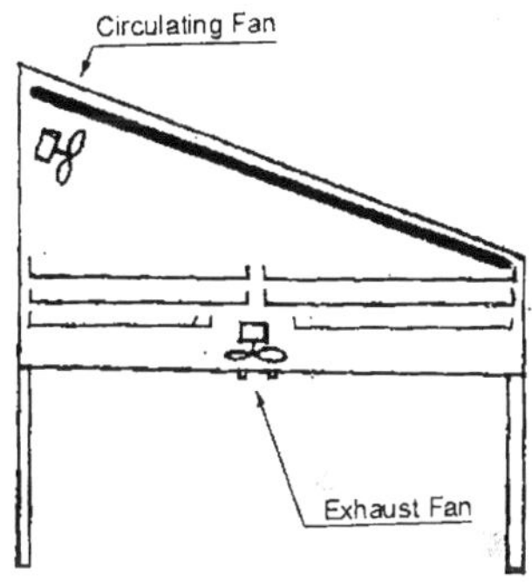

Figure 1. Starcken model.

In the late 1970's I built a cardboard and plywood dryer from plans in the popular California-based magazine Sunset - see fig 2. The only modification which really seemed necessary was to shield the food from direct radiation. In the original design there was a loss of chlorophyll when drying herbs. Simply placing a piece of cloth or paper over the window of the dryer kept the herbs green.

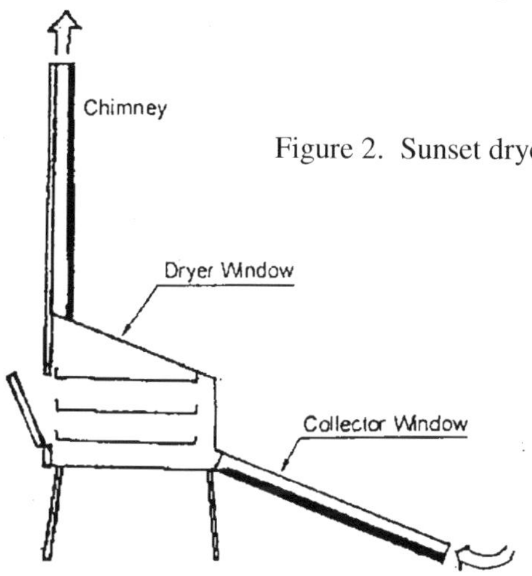

Figure 2. Sunset dryer.

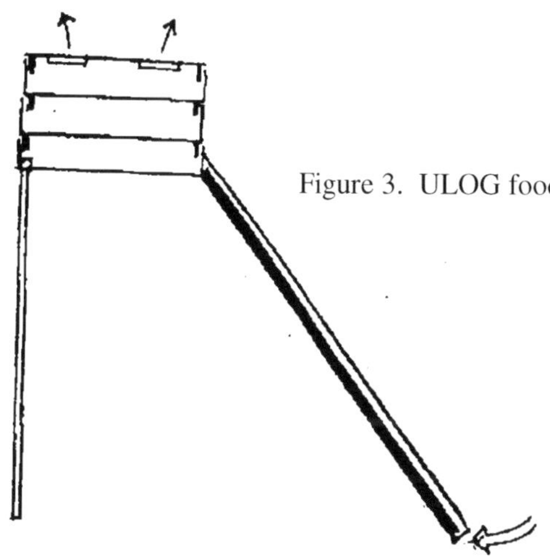

Figure 3. ULOG food dryer.

For some time the ULOG (Swiss) designed dryer has been popular, but one inconvenience of the ULOG dryer is that in order to check the foods drying progress one has to disassemble the top part of the dryer. The simplified drawing in figure 3 is of the ULOG model.

The Barker cabinet dryer consists of expanded metal mesh trays in an aluminium cabinet surrounded by corrugated polycarbonate and with openings top and bottom to allow manually-controlled air flow - see fig 4.

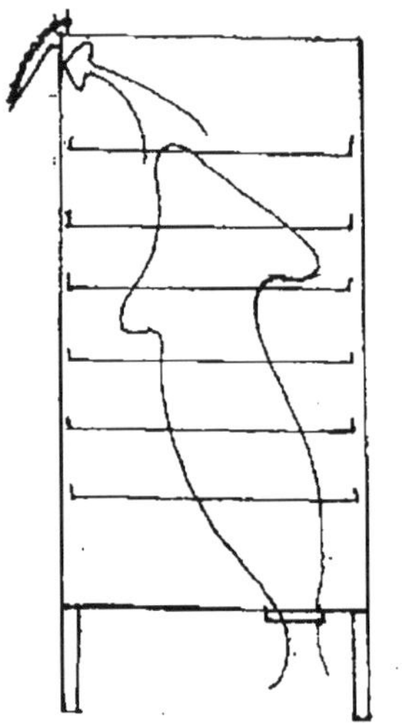

Figure 4. Barker model.

Drying takes from a few hours for herbs, to a few days for bananas and pears. It very much depends on the thickness of the slices and the dryness of the surrounding air.

Books advise that food is sufficiently dry when it has 85% of the moisture removed. The average householder may be unable to work that out, but let's just say dry till the food is similar in feel to commercially produced food from your bulk food shop.

We have dried apricots, apples, pears, banana, peaches, figs, tomato, onion, chillies, cauliflower, eggplant, capsicum, mushroom and many herbs such as mint, parsley, and lemon grass.

Dried vegetables are then available out of season for addition to, for example, winter minestrone.

There are some rather exact procedures for pre-treatment (blanching, using preservatives such as sulphur dioxide and sodium nitrate) and for proper storage, but at home you can often wing it and get good results.

Drying is processing so there are inevitable nutrient losses, though not nearly as much as, for example, in canning.

It is important to remember that most dried foods available from a supermarket are processed in commercial dryers using fossil fuel. In this information-rich age when growing, cooking and repairing skills are fast disappearing it is very satisfying to take that small, but significant, step to reduce greenhouse gasses and produce your own dried foods.

There are no hard and fast rules for constructing a dryer. Good plans are available from a variety of sources, including the internet.

Alternative Technology - Large-scale Chicken Tractor

Alternative technology (often called soft, intermediate or radical technology) is described as "Technology which is soft, is gentle on its surroundings, responds to it, incorporates it, feeds it. A nuclear power plant doesn't qualify." It is a combination of alternative means of energy production (e.g. water, wind, solar, biofuels, pedal power) and energy conservation. This latter aspect is very important as it often causes us to re-examine and re-design systems in a more energy-efficient way.

The importance of insulation and draught control in "solar" homes is an example of such thinking. The technological possibilities of plants and animals are incorporated. For instance, windbreaks to channel wind onto a generator, deciduous trees for seasonal shading, and the chicken tractor are all basically technological ideas using plants and animals for their implementation. I believe that alternative technology offers a more sensible and sustainable system of energy manipulation than our present system and because of these reasons is more suited to permaculture.

It takes time, money, space and, above all, ideas to develop alternative technology and perhaps it is because of these constraints that very little is happening (or seems to be happening) in our society at the present time.

I was reminded of the permaculture principle "every unit has many uses; every use is achieved in many ways" because the same kind of philosophy is inherent in alternative technology. Likewise, the solutions which alternative technology offers to technical problems are similar to the solutions permaculture offers agricultural problems: a diverse range of sustainable (renewable) units, intelligently arranged to obtain the most productive and permanent system possible.

As an example, let us examine the design of a particular chicken tractor. The chicken tractor is basically a rectangle of land 13 m by 42 m divided into 12 seven-metre squares with a 1 m wide chicken run down the middle. The chicken run and a total of 20 frames were made from angle iron and chicken wire, and took three month's work. These frames are assembled to make 2 pens (each 7 m), one for each side of the chicken run, which can be shifted by two people in a few minutes.

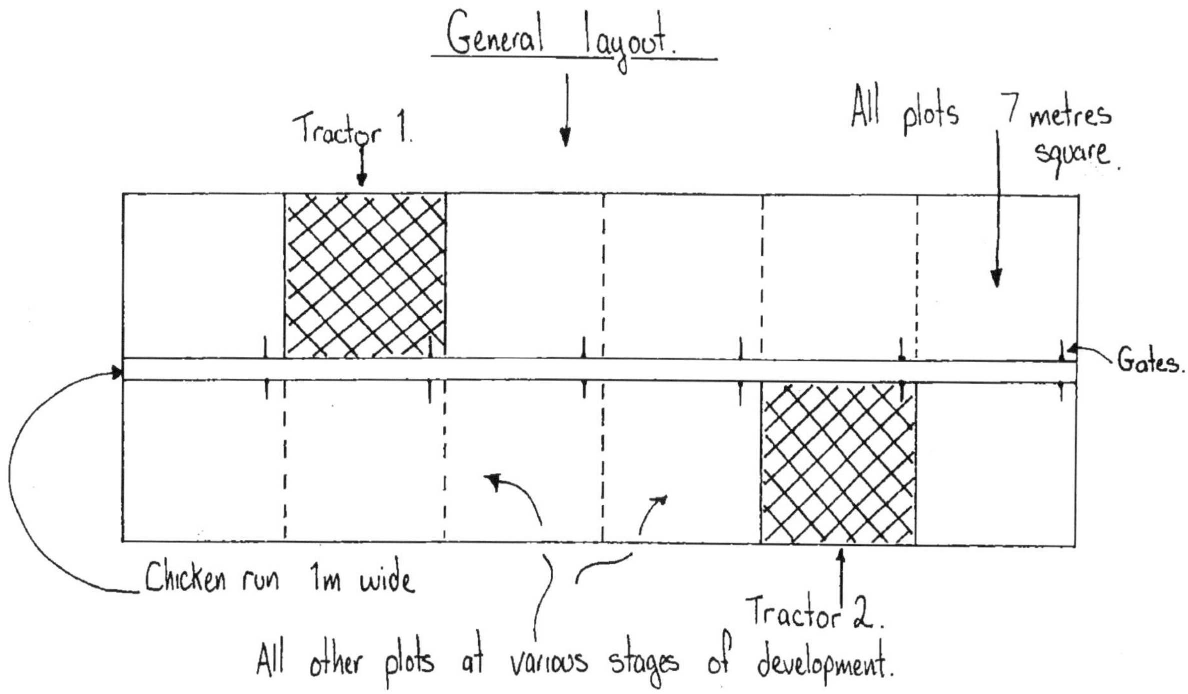

The main chicken run has gates to allow access to each of the pens so no matter where the pens are, the chickens have access to both. The posts which support the lid of the tractor also serve as supports for plants. The whole area is trickle irrigated. The principle of operation of a chicken tractor, as you probably know, is to keep the chickens in an area to clear it of weeds, aerate the soil and manure it (while producing eggs). Once this is done the pens are moved and vegetables are planted on the newly ploughed plot, while the chickens continue to work elsewhere.

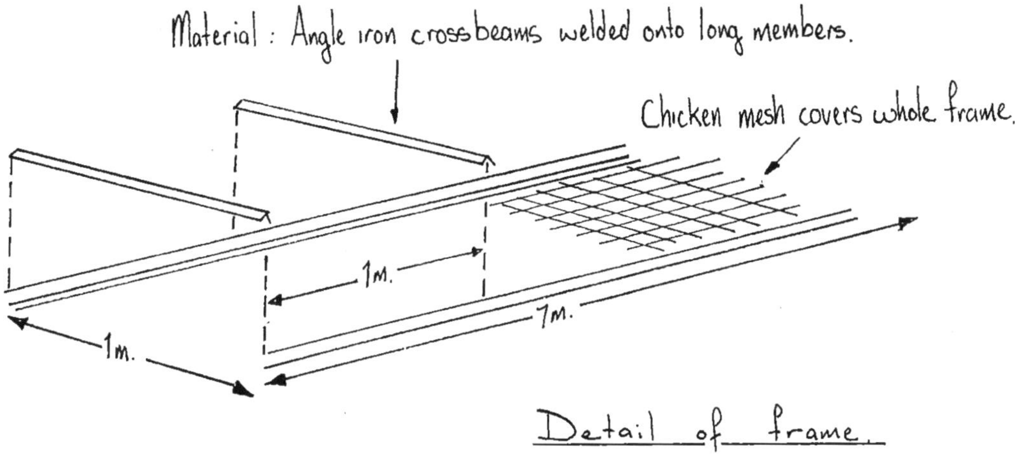

This design is a double tractor - one on each side of the chicken run, and each having 6 plots. The chickens can clear the land in two to three weeks, so the overall cycle time is 12 to 18 weeks - ideally suited to growing vegetables.

The results were impressive. The vegetables were big and healthy and the chickens were active and healthy (and producing a lot of both eggs and new chickens). Small chickens need to be protected from crows and other birds of prey - the chicken wire they used was big enough to let the chicks out where they fell prey to the birds. Around the bottom smaller wire size could be used to prevent this from happening.

Proposed extensions are the addition of a glasshouse at one end (to utilise the chickens as heaters and a CO_2 source) and the planting of perennials around the general area. All in all, this chicken tractor is the most well designed and productive 500 m^2 (its total area) I've ever seen.

Notes on Fruit Drying

The direct rays of the sun, especially on really hot days, can be used to dry fruit. We have used a home-made fruit dryer box and hot air tunnel before, but have found that it is less time-consuming to do it in direct sunlight.

Using a piece of welding mesh as a rigid base, we spread a fruit tree bird net over the top and fruit cut into suitable pieces over that. Cover with bird net - this seems to deflect the attentions of most of the bees and flies too.

To avoid ants, we stood the whole thing on 4 blocks of wood which stood in plastic icecream containers of water. As the fruit dried, we turned each piece over, repeating almost daily, until the fruit was dry enough to store in a glass jar.

We don't use any preservative and good judgement is needed when the fruit is sufficiently dry. The apricots will look nothing like those in the health shops but will respond to soaking in water. This past summer was so hot we tried drying mulberries with great success, placing them in direct sun on baking trays and giving them a daily stir. Figs too are easily dried by cutting into about 3 pieces.

As an experiment we've tried drying carrots, beans and tomatoes too. Grate carrots and place on baking trays. Slice beans as for cooking, also slice tomatoes and dry similarly. Store in screw-top jars when perfectly dry.

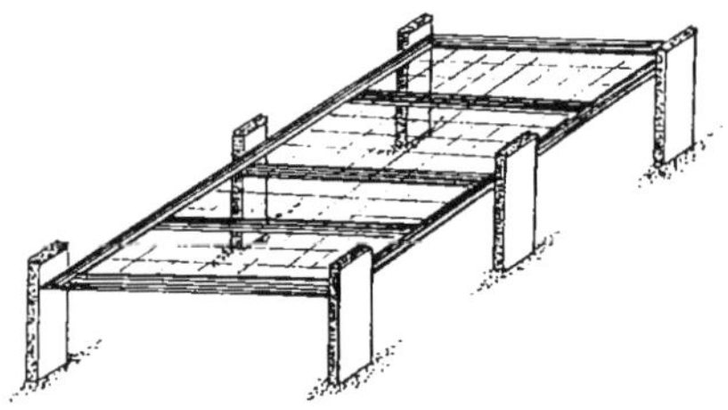

We use the fruit mainly in fruit loaves, after soaking overnight in a little water, adding perhaps sultanas and lemon peel.

Irrigation

Irrigation: adding water to the soil by whatever means one believes is best for their situation.

We will look at various methods which have been used and proved successful by many people over the years. These include:

1. Hand watering.
2. Watering by hose and sprinkler.
3. Trickle irrigation.
4. A complete sprinkler system with underground pipes.

Hand watering may be liked by the more conservation-minded, but hand watering with a hose can be very frustrating when it has to be dragged round shrubs and trees. In a small garden I would suggest a 10 litre plastic watering can as the cheapest and most economical way to water a few native trees and shrubs. Plant the area out at the beginning of winter and mulch round the root area leaving a small depression. Five litres of water twice a week for the first summer poured into this "basin" should be enough to get the plants going and from then on they will fend for themselves.

A hand-held hose can be used for a small grassed area bearing in mind that it is far better to run a hose 15 minutes where you need it than use it for 30 minutes with a 50% waste factor.

Many small gardens have vegetable beds - raised or otherwise. These can be watered by either can or hose. If watering with a hose, stick to a given length of time - trial and error will determine just how long. Mulch the garden beds to increase the value of the water you do use.

Watering by hose and sprinklers would be one of the most common methods used in the garden area. Put the hose on the tap with a sprinkler on the end, turn on and forget it! Very wasteful. If using this method, there are a few things we must bear in mind. One sprinkler head delivering a full circle is rather hit and miss on its own. It is now possible to get a variety of heads which will place the water where it is needed. Another very good idea is to have a time clock with this method so that you can set and forget it. The water can be turned on any time, when going to work, going to bed or sitting down for a meal and it turns off automatically. I found this a very useful way of getting my watering done without too much effort. One hint about the timer is to have approximately 1.25 metres of hose connected permanently to prevent continuously plugging into the clock as this tends to break a good many fittings on the tap. Timers will come apart to clean the pin which controls the cut out, so don't throw away without checking if it ceases to work.

The Trickle System is very economical and quite simple to install. Follow the rules and you will put the water where you need it. It is easy to get a leaflet from the suppliers of this type of equipment which will explain all. In the case of native trees and shrubs drip feed can be used for one or two seasons until the plants are established, and then the system can be moved to another area. On the average block this system is used off the mains water supply.

A complete sprinkler system can also be connected to the mains water supply with several stations controlled by low voltage automatic switches on a time clock. I find this a bit wasteful as the same amount of water is used whether it is 25°C or 40°C so I favour a certain amount of manual control.

When putting in this type of system future flexibility is the name of the game. Allow for removing sprinkler heads and capping. Fix valves where a hose can be fixed to put a little extra water in that odd corner or behind the bush that grew so quickly so no water reaches the plants behind it.

I must conclude by saying that under-watering is foolish because when we lose plants and vegetables through lack of moisture, all our water, time and effort has been wasted. Equally, when we over-water a great deal of water is wasted.

There are a good many different systems being used by different people very effectively. I can easily claim they cover almost every type and variation possible. However, when watering a given area we must endeavour to use that area to its utmost potential.

Raised Beds for Vegetables

First select a sunny position. Work out a plan for the area you have available, bearing in mind that your beds need to be approximately 1 to 1.5 m wide and any length, to suit space available. Multiples of beds must be far enough apart to get between them with a wheelbarrow to put on more growing medium at a later date.

Beds can be started quite easily with a frame 0.5 m deep. For this, old sleepers, bricks, concrete blocks and so on can be used. Put down newspapers about 0.6 to 1 cm thick and then put down the surround. Two sleepers 2 m long will make a bed long enough to start, and one trailer of manure will fill this size.

Usually the horse manure available contains enough sand to make it suitable for immediate planting. Make sure it is not 'fresh' as this will compost down and produce much heat, killing plants. I have planted a bed like this and harvested two or three crops in quick succession before adding any more material, when the bed had receded to 10 cm thick.

Some of the many advantages of the raised bed are no digging, because we don't tread on the soil to compact it; the material we grow in is much healthier than normal soil, so can be planted in an intensive or close spacing way; it is more economical on water - twenty litres (2 laundry buckets full) a day will keep a bed of the suggested size in good condition through the worst heatwave. It is easy to construct a small frame, which can be placed over the bed when shade is needed during a heatwave.

Electric Fences

The availability of solar cells and good batteries make the use of electric fences a viable option. With a bit of care, building an electric fence rather than a more solid, 'normal' one should be considerably cheaper. Insulators can be made of lengths of PVC, slid over the live wire and attached as in diagram 1.

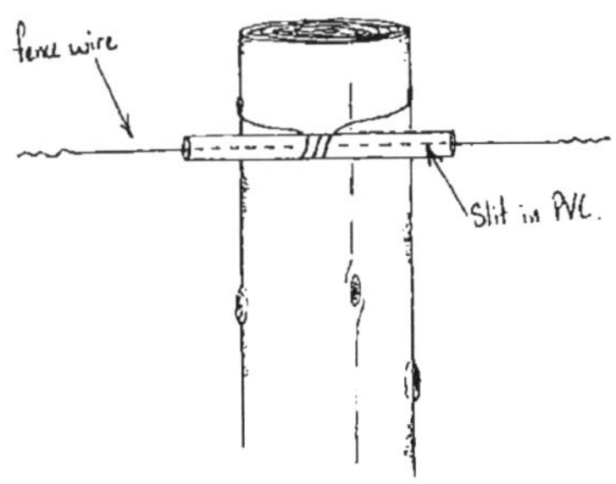

Diagram 1.

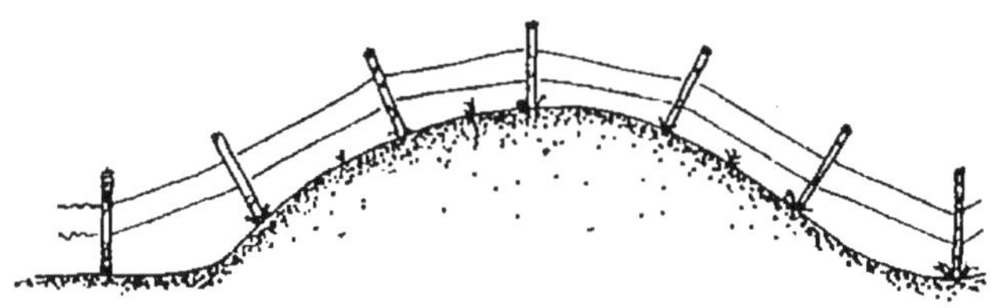

Diagram 2.

Great care must be taken to ensure the wire holding the pipe does not touch the live fence wire. I have seen a set up where the post had a hole drilled through it and the plastic inserted through the post. An electric fence uses less upright posts than a conventional fence as it is the "fright" value of the shock which is used to control the animals and not the strength of the structure.

If you are lucky and live on flat ground then the posts can be well spaced out. If the fence is on rolling country, the spacing will vary to suit the lie of the ground as in diagram 2.

The wire can be of a lighter gauge than conventional fence wire and of a lower quality, as less strength is required. You may even use mesh if you are trying to control small animals or to stop larger ones from reaching through the fence. Barbed wire might even be suitable, but both mesh and barbed wire may give a lot of radio and TV interference: not something to endear you with the neighbours. In the same way, the posts themselves can be lightweight poles, rather than heavy fence posts.

However, it would probably be wise to put in a good-sized strainer and corner posts. Because the tension is less than a normal fence, you could use a standing tree, which will suffer little damage if you cover the wires passing around it. To apply tension to the wires you will need some form of strainer (see diagram 3).

The main problem with this will be the gap between the strainer post and the batten, which can be filled with either wires running down from the top or light sticks attached to both wires.

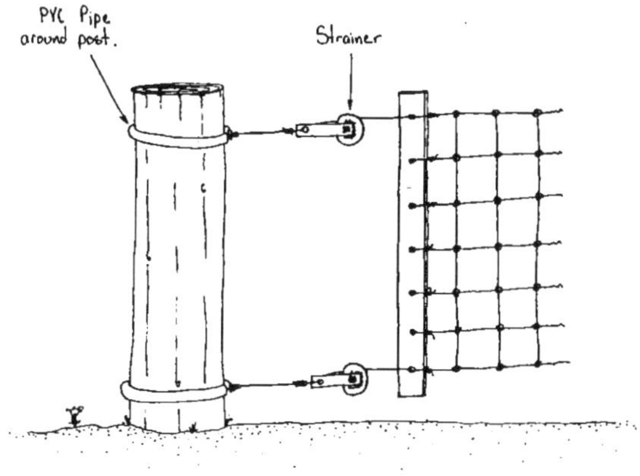

Diagram 3.

The disadvantages are the initial, one-off cost of the unit, which should last you a long time. You might consider an 'unshortable' fence, as they save the need for clearing away growth by the fence (although this type of maintenance will also improve the effectiveness of an unshortable unit). With normal systems, any tree branch or grass in contact with the wire will short the fence and make it useless, whereas, an unshortable fence will also charge the ground where the wire touches, and thereby give the animal a shock.

Another cost is the battery which should remain nearly fully-charged if connected to some form of charging device such as a windmill or solar panel system. Diagram 4 shows how to make a simple tester for your electric fence.

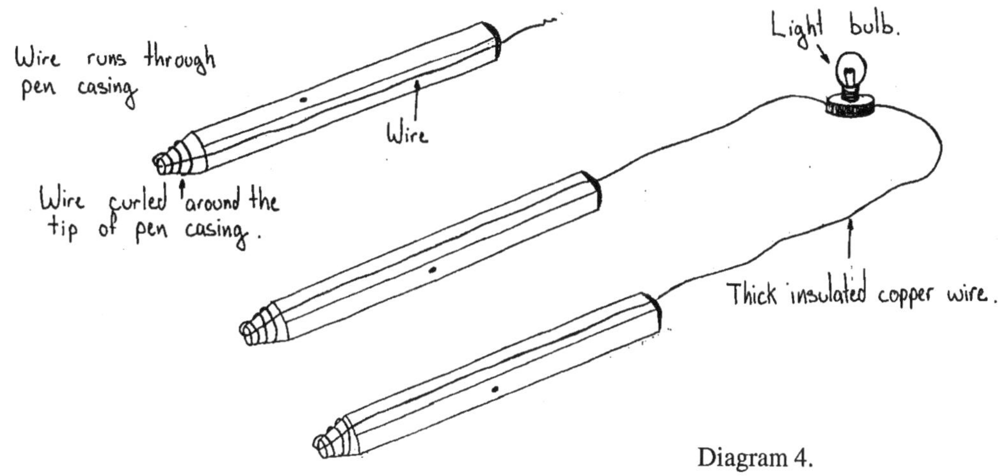

Diagram 4.

Sulphuring Fruit Before Drying

Apricots, nectarines and peaches that have been sulphured before being put out to dry in the sun retain their colour really well and can turn out as well as the shop-bought product. Sulphuring ensures better flavour and less loss of vitamins A and C. The fruit is also less likely to sour, get mouldy or be attacked by insects during drying or storage.

For the sulphuring box you need a large, airtight wooden box, such as a tea chest or large cardboard carton. I use a lightweight, durable box made of scraps of plywood. The joins have to be taped to prevent the sulphur fumes from escaping.

The trays are made with slats 1 to 2 cm apart with 10 to 12 cm high sides so they will be well spaced when stacked. They need to be made of wood not metal, but preferably not pine as it is said to give a resinous taste to the fruit. I used a tuna tin to burn the sulphur in (one teaspoon sulphur for every half a kilo of fruit.)

Steam blanching fruit halves for 15 minutes before sulphuring discourages oxidation and is used for peaches which go brown quickly, and for removing their skins. It is also used for cracking the skins of grapes, prunes and figs before sun drying.

Times for leaving halved fruit in the sulphur box are 90 minutes if blanched and two hours if not blanched.

The fruit halves are put in single layers in the trays, cut side up. When sulphuring is finished, they are full of juice and the trays are shifted out into the sun. The fruit dries more quickly on a hot tin roof, or if covered with a piece of glass for the first two days. I rest the trays on boards so there is air movement underneath and through the slats. A solar dryer would work on dry, sunny days, but in winter you may have to dry pears and apple slices on trays standing on a metal rack on a stove.

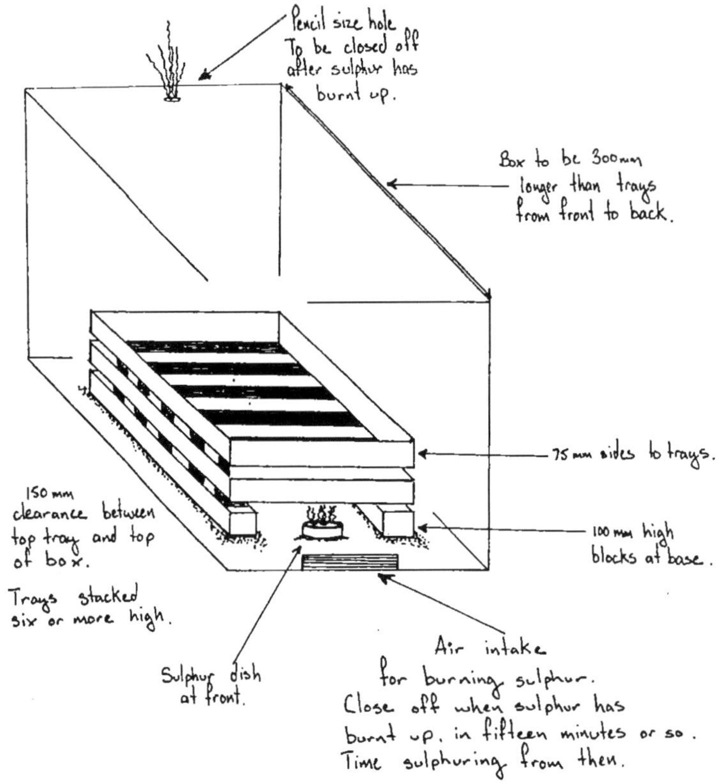

Building a Small Grain Mill

A reasonably efficient mill can be manufactured from readily available materials by any competent handyman or tradesman.

The basic requirements are a readily available ordinary carborundum grinding wheel, and a grinding-sledge which you would have to manufacture yourself. You can make the contraption wider by fitting two or more wheels next to each other, as long as you don't forget that their weight may demand a stronger shaft and that you will need more power to drive it.

After you have decided how big you want your mill, make up a strong framework to support the bearings of the wheel, the pivot-bearings of the sledge, the sledge adjusting mechanism and all the oddments that you need to feed in the grain and to collect the meal, and a cover to stop the dust from suffocating you. You could use a small vacuum-cleaner to create minimum pressure inside that cover. Or you could even use a vacuum-cleaner with a disposable bag to collect both meal and dust. I will leave those things to your own ingenuity.

For the grinding-wheel shaft you could use thick-walled piping with a tight-fitting smaller rod as the core. At one end of the shaft weld a flange, at the other end cut a thread sufficiently strong to clamp the grinding-wheel in place. The solid core should be long enough to accommodate good bearings at either end, to carry the power-transfer (V-belt or chain) and to allow for the cover, which should fit tight to the wheel and sledge to prevent the grain by-passing the system.

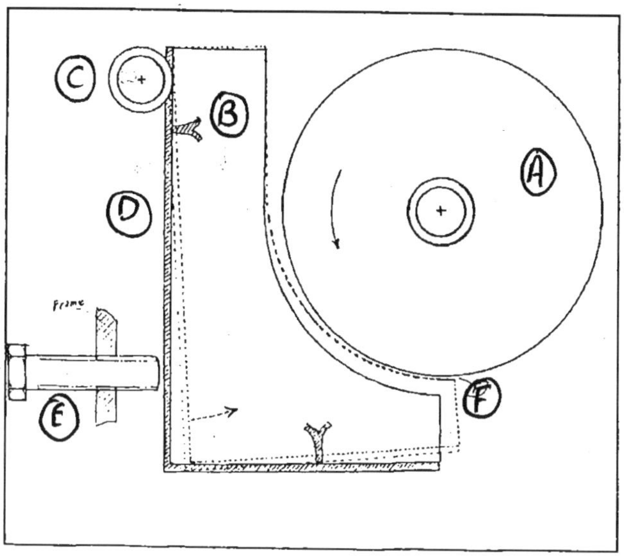

On a scale of 1:2.5 it represents a grinding wheel of 125 mm radius and a parallel gap of 15 mm between wheel and sledge. When pivoting the sledge around (C) the shape of sledge and wheel form a feed-in gap for the grain. If you want a bigger gap you can change the shape of the sledge and slant the vertical edge further back towards (C).

In this basic form:

(A) is the milling wheel;

(B) is the grinding-sledge;

(C) is the pivoting point for (B/D);

(D) is a metal cradle for (B) made of 6 mm flat-iron or U-profile;

(E) is an adjusting mechanism to reduce or enlarge the milling-gap (F).

To manufacture the grinding sledge you can use different approaches. The sledge will have to be the same thickness as the milling wheel and it needs to be firmly attached to the metal cradle.

The best way of doing this would be with anchors welded to (or protruding through) the cradle and extending into the sledge material. High tensile material might be preferable if the sledge is not very thick, so as not to have to weaken the concrete too much.

The radius of the milling gap should be slightly larger than that of the wheel; on the sketch I have allowed for another 10%. Building a mould should cause no problems and I won't enlarge on that.

Your best shot would probably be to use blue metal road dust for concrete aggregate but, because of the rather small contact area at the milling-edge, you shouldn't use large chips of blue metal.

So screen it to remove anything bigger than 3 to 4 mm and then screen it again through fly-screen gauze to remove all dust. I would then put this aggregate in a container and pour in a measured amount of water to see how much is needed to fill all gaps in that coarse material. Drain the water out and use the same quantity of cement to make a batch of 'concrete' of normal consistency. Pour this into the prepared mould and let it set for a week or so.

Assemble the bits and pieces and you should be in business. The speed of the grinding-stone should be no more than 50 metres per second at the outer rim (which amounts to just over 3000 rpm for a 30 cm diameter wheel). Anyhow, keep the speed below the rated capacity for the industrial use of the grinding-wheel.

The drive is your own choice, depending on how big you made the whole thing: you can use a geared-up hand-drive, a push-bike or a washing-machine motor.

When building something like this it is a matter of common sense what materials to use but if you are going to err it is better to err towards strength; the whole thing should be strong enough to last. You'll soon learn how much grain you can feed in; the mill will choke up if you overdo it.

Plant Support Systems

Fences and structures are important for the support of many of our common permaculture plants. These plants include Grape, Youngberry, Boysenberry and Passionfruit. The first three will grow quite happily on post and fence wire construction which can be 1.25 m high with two other intermediate wires. The passionfruit however, would be much better on a fence made of posts and old chicken netting or similar.

With all the various fences one must consider the overall environment, such as where the wind and sun come on to the scene, e.g. a long fence with solid growth may suffer badly from a gale force wind. Therefore it is better to make a fence either Y shape, U shape or V shape. Another advantage with shapes gives us a slight variation in maturity of crop.

Grapes and passionfruit can be grown on a wall or fence line, e.g. a picket fence. The berries are a crop which are better harvested and maintained from either side.

Alternate material for all these fences can be concrete steel mesh. Off-cuts of this can be obtained from building sites or can be purchased from building supply hardware stores in sheets, usually 6 x 2.4 m. When buying whole sheets it is possible to roll them into an arch giving shade underneath to grow suitable crops when the arch is covered.

Climbing vegetable crops include beans, peas, chokos. The first two can be grown on a chicken wire or stick fence two metres high. With chicken wire, use an upright stake at one metre intervals. I do not recommend wire as a first choice because a gale force wind will devastate the crop in an open position. With a stick structure beans and peas wind round the sticks making a more stable situation and reducing any risk of wind damage. If you only have materials for a low fence for runner beans it does not matter as you just pinch the ends off the beans when they grow above the structure.

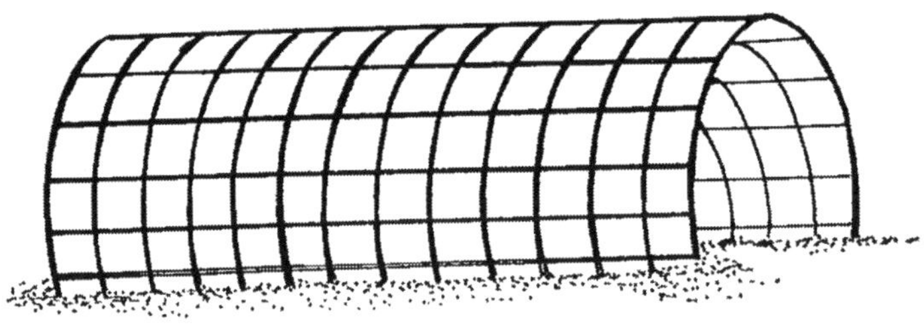

The choko needs a fairly heavy structure or it will run up a few sticks and find its own support in the nearest tree making it very hard to harvest with or without a large ladder. So try to grow the choko on a solid frame such as the re-inforcing steel mesh. This can be erected in various shapes as for fruit crops.

For all our climbers we needn't purchase anything new in most situations. Watch when the neighbours throw out that old rusty gate or put up a new wire fence. These cast-offs are all climbing supports and the posts will support any old piece of chicken wire.

General Handy Hints

Canola for chooks

When sowing green manure crops (broad beans, rye corn and lupins) in March, sow a patch of canola. It makes an excellent cover crop and germinates quickly and thickly to keep weeds at bay. Bulk seed may be obtained in small quantities from grain stores. Canola can be used as greens for chook food and is best sown in Autumn.

Eggshell powder as a seed helper

Save eggshells, dry them thoroughly (on the woodstove, in an oven after a meal is cooked, in the sun), crush them into a fine powder and store in a convenient dispenser such as a metal spice box.

When you sow seeds directly into the garden, sprinkle the eggshell powder over the seed site to mark the site of the seed, to add calcium to the soil and to deter soft-bodied pests such as snails and slugs.

Seed saving beetroot and carrots

Both of these vegetables give large quantities of seed from even one plant, but have a long lead time to when you actually collect the seeds. Select two or three plants, leave in the ground and in due course they should show signs of going to seed by means of a branching shoot from the centre of the plant.

Over a period of months this will grow very tall (up to two metres), so early on place a couple of stakes around the plant so branches can be tied with some sort of stretch fabric.

When beetroot shoots have died off and carrot flowers have dried out, pick the plant and leave it to dry thoroughly before removing the seeds. This process can take nine months or more from the original sowing.

Drainage for pot plants

Rocks, stones or broken crockery in the bottom of your pots are essential for good drainage. However, these substances can often make larger pots heavy and difficult to turn or move. Try using broken up polystyrene foam from old seedling punnets or packaging as a light, and just as effective, alternative!

Topsoil

Weeds usually create good soil that hangs on their roots. Cut the weeds off (with a knife) just below ground, or mow them, and the roots rot well in the soil, without drying the soil out by pulling them out. Or mulch them to death.

Growing peanuts

New gardeners - try a small plot of peanuts. Buy raw peanuts from the local health shop. We sowed 40 seeds in November, 10 to a row about 15 cm apart in the row, making a small block of ground to look after.

We dug them in early May and were amazed at the quantity of full shells on the root structure. Leave exposed to dry for a few days, pick and shell, roasting carefully in a moderate oven or grinding and adding to muesli biscuits.

Green manuring - plus millet for chook food

Don't leave a bare patch of soil anywhere in your garden - make it work for you. You will grow weeds anyway so why not grow something which will add nitrogen to the soil.

During winter, broad beans will grow anywhere, anytime, practically without any assistance and even if you don't eat them the plants form a great green manure. Seeds can be planted through to August.

The root system of each plant will produce nitrogen nodules. Try not to pull up the plants, roots and all, but scythe or mow down leaving roots in the soil.

This resulted in a good mulch cover into which we sowed millet seeds, harvesting these for chicken feed.

Elder flower champagne

Pick 6 Elder flower heads. Shake out any insects. Place in a large clean bucket, add two slices of lemon, 1 kg white sugar, 2 tablespoons white vinegar and 4.5 litres of water. Mix well. Cover with a clean tea towel and leave for 24 hours. Strain and pour into screw top bottles. Seal. Leave for 2 months to brew but check every few days to release any excess gas that might cause the bottle to burst. Wine is usually produced after a few weeks, but it is best to leave it for longer.

Other helpful hints

- Store pumpkins on a shed roof. If left in a shady place, they don't keep well.
- Pumpkin (raw) can be frozen indefinitely if blanched first.
- To ripen home grown bananas, pick green, after they change from a miserable square shape to a nice fattened-up shape. Place a layer in a newspaper lined box with a few Granny Smith apples. Cover with newspaper and continue the layers. The apples give off the ethylene gas which ripens the bananas.
- Comfrey leaves are great as green manure particularly under potatoes as they are planted out.
- Two points on mulching were: if the manure used is smelly and upsetting the neighbours, just cover it over with seaweed. It works like a charm. Secondly, don’t cover potato crops with seaweed - it lets the light in and greens the spuds.

Wonderberry jelly

6 green apples roughly chopped (and peel and pips)

1 cup *Solanum nigrum* berries (Black nightshade)

Cover with water and simmer until the apples are soft. Put into jelly bag or flour sack and let drip over a bowl overnight. Do not squeeze. Feed pulp to stock animals and measure the remaining liquid. Add one cup sugar for each cup of juice. Add a packet of jamsetter with sugar and boil for 15 minutes. Let sit for 10 mins. Skim if necessary and bottle in hot sterilised jars.

To Sludge or Not to Sludge?

Introduction

The sewerage system uses water as a transport medium. Just as we use bikes and cars to get to places in the city, the sewerage system relies on water to carry the waste to the treatment plant. Water constitutes 99.9% (with dissolved salts) of the material that the waste facility receives. Only 0.1% are solids. That figure is enough to make the ignorant like me think.

Western Australia produces about 200,000000 litres of wastewater daily. The Subiaco Treatment Plant takes in for treatment 50 million litres of waste per day from 300 000 households in an area encompassing Scarborough, North Fremantle and Guildford. 85% of waste is sourced from domestic points, the remaining 15% from industry under strict controls.

Each person in Perth sends (on average) 180 litres of water per day down the sewer, literally. This is mainly sourced from showers, toilet flushes, washing machines and dishwashers. The more affluent the society is, the more water waste is produced. Dishwashers and their rinse cycles were the worst offenders.

Since Perth has sandy soil which can take septic tanks, only some of Perth metropolitan homes are sewered. Most other cities which lie on a different geology have some 95% of homes connected to sewerage systems. As we now know, to connect Perth’s 300,000 unsewered homes requires 800 million dollars. The simple sums says that would cost each connection $2670. But the reported connection cost is $1670. Who pays for the difference? The water contains nitrogen at 50 milligrams per litre (mg/l) and phosphorus at 10 mg/l. The treatment plant recovers approximately 50% of that in solid form - that is, 1.3 tonnes of high grade nitrogen per day.

The process

There are four treatment processes: Preliminary treatment, primary treatment, secondary treatment and sludge digestion.

Preliminary treatment

Preliminary treatment consists of adding chlorine to the water at the point it enters the plant. The chlorine is used to control odour and costs $1500 per day. The water then passes through screens to collect large solid items. Some of these were on display: toys, cotton buds, syringes (and they weren't small ones), jewellery and a host of surprising items.

Primary treatment

The water then passes to the primary treatment area which consists of a settling tank to separate the grease (that floats), the solids (that fall as sludge) and the water. The water stays here for two hours and the grease is skimmed and sent to a high temperature incinerator at the preliminary treatment building. The solids are transported to the digestion process.

Secondary treatment

In the secondary treatment, the water is treated by biological processes. Subiaco uses an activated sludge process where naturally occurring bacteria feed on the organic matter. The process is accelerated by blowing oxygen (air) into the effluent to further multiply the bacteria (activated) which in turn accelerates the process of converting organic matter into solid form. This takes 8 hours. The water is passed into more settling tanks where the sludge is separated. At this point the water is quite clean and very clear.

Some of the sludge (which contain the bacteria) is returned to the activated sludge process. The remainder is taken to the digestion tanks and the water is released into the ocean.

Sludge digestion

The sludge from the primary and secondary treatments is collected and de-watered by squeezing it between two continuous conveyor belts through a set of rollers. The resultant is a damp cake, looking like squeezed and blended compost. The sludge digester is a sealed tank and uses an anaerobic process to digest the bacteria. It is kept at 35°C for 15 days. Apparently, the undigested sludge contains some life that is not exactly benevolent. Amongst other animals, hookworm and intestinal worms populate the sludge.

Discussion and issues

Historically, sewerage systems were an engineered solution to an endemic problem that governments could not understand. London built its sewerage system because of disease and polluted water supplies. At that stage bacteria were unknown. The epidemiologists (public health researchers) and engineers knew that cess pits near water wells produced cholera and typhoid. Sewerage systems were the solution to what was an urban problem - too many people living in a confined area. In the 1990's with water resource management an issue and alternative treatment techniques known, perhaps another approach could be taken.

The 180 litres of waste water for each Perth resident represents only 60% of the total drinkable water consumed. The remaining 40% goes onto the garden - 120 litres per day. If we could divert 120 litres of waste (as grey water) onto the garden, reduce fertiliser consumption and change our toilet system, then we would not be wasting any water. But can we expect every person to take a responsible attitude to their own waste water and human waste?

The reliance on technological solutions was pointed out when the south west power failure occurred earlier. Untreated sewage was flowing into the Swan River unchecked. The need for constant power to keep our cities clean must mean we are riding on a pinnacle - an upside down triangle. Is this really the way we want to live? Permaculture can provide a set of alternatives so that we need not rely on just one. Can we design redundancy into our homes to make them resilient for when things go wrong? Can we measure our affluence in this way instead of dishwashers? Measure the amount of urine you produce, then measure the flush. You are left asking why?

Conclusion

For historical and cultural reasons, wastewater treatment has been assigned as a public activity - that is, performed by instrumentalities set up by government. Over time, people have empowered government to dispose of waste produced by themselves and industry.

Water treatment seems to be effectively handled and a lot of thought and planning has gone into the treatment of waste by public officials.

The need for automation (removing grit and de-watering) and the necessity of concentrating the natural processes into a small area, has produced a pipeline system and plant drawing a large capital investment. The large concentration of people in a small area requires massive public investment of capital.

The question is whether people ought to take up the responsibility of treating their own effluent. Just as people are beginning to outlay discretionary funds on rainwater tanks, should people also spend resources on household effluent treatment? It is up to each of us to be responsible for the impact we have on our environment. By empowering ourselves with that responsibility, we will learn what that impact really is.

There are many techniques available to individuals to reduce the water sent to the treatment plant. These techniques also reduce the demand for new water resources. By picking up this challenge we can affect much more than we have changed.

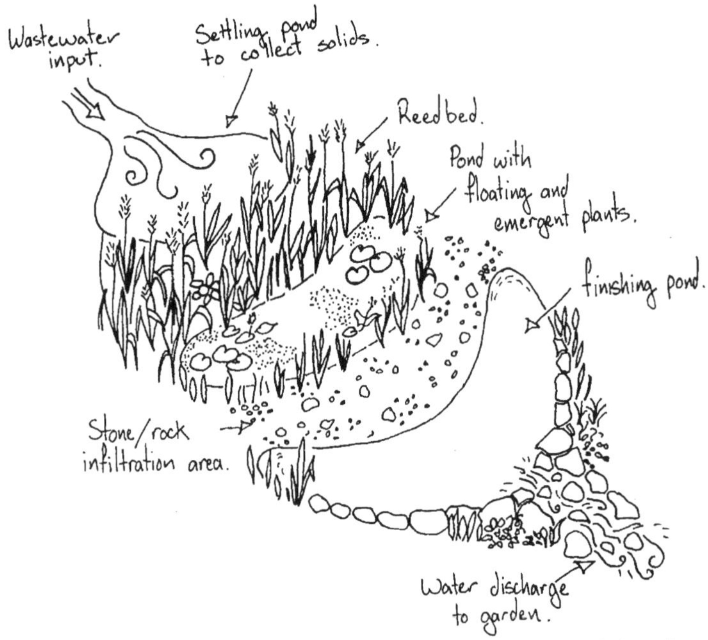

"These MARRON are so tasty! ... pity we can't get them to grow BIGGER"

Distributing Greywater

The best pipes for distributing grey water are polythene (black and flexible) and PVC (stormwater pipe). Both come in different sizes and the smallest useful size is about twenty five millimetres.

In some cases the distributor can be attached directly to the drain plug on the "S" bend sludge trap but is usually not that simple.

Diagram 1 shows how I drain my shower water. Drilling a hole through concrete is difficult. I used a hammer and cold chisel. An impact drill would be better and a jackhammer best.

Diagram 2 shows how I drain my sink and wash-trough water. Making a watertight valve is difficult. A rubber solution helps.

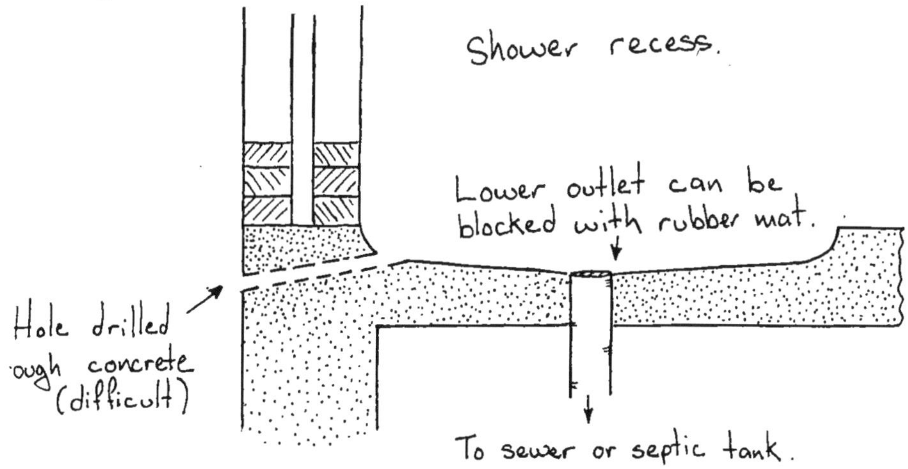

Diagram 1.

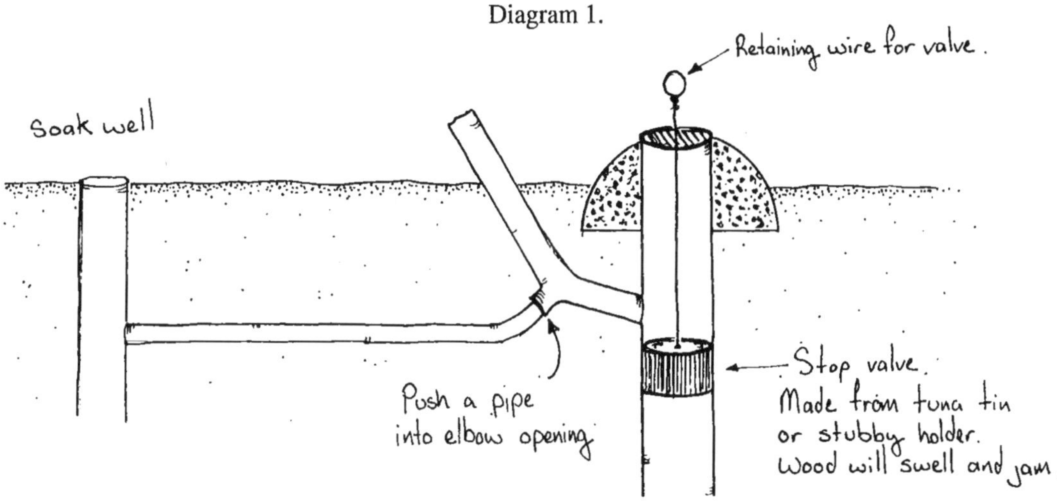

Diagram 2.

Water Conservation

There are heaps of good water conservation hints. Here are a few of them!

- Do without a dishwasher (or use only when full).
- Choose a washing machine for minimum water use.

- Install "Flickmaster" taps (these are fully on after a quarter turn).
- Install a showerhead which reduces water use.
- Install a dual flush cistern in the toilet.
- Locate your hot water system close to the point of use, or have special on-site heating systems as used in camping.
- Install a rainwater tank for all uses other than drinking. Such a tank would collect water for garden use.
- Pipe your greywater to the garden. (Greywater is what flows out of the bathroom, laundry or kitchen sink after use). The large volume gained this way would be sufficient to keep the garden watered. A drainage system which filters greywater through a grease trap or sand filter would be desirable. Another way is to direct the water into subsoil drainage like a leach drain from a septic tank. Or, use evapotranspiration techniques - lay drainage pipes on black plastic, cover with mounds of soil and plant on top.
- Find other ways of dealing with human waste! A waterless loo would allow the use of faeces as a fertiliser. If dried or composted and allowed to reach sufficiently high temperatures for only four hours, pathogens are destroyed. To be safe, the compost should be kept for six months and covered with soil when used.

As a point of comparison, the pathogens in faeces transported in water (as in our sewage system) live for 6 months.

The 380 L of urine produced per person per year could be used to fertilise fruit trees. Simply collect, add lime, stand for a day and dilute with water. Alternatively, add urine to a container of sawdust and use this on the trees. Best not to use on vegetables in case of pathogens.

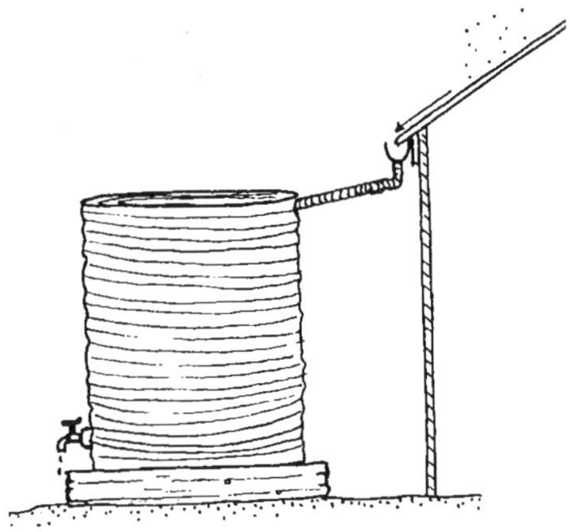

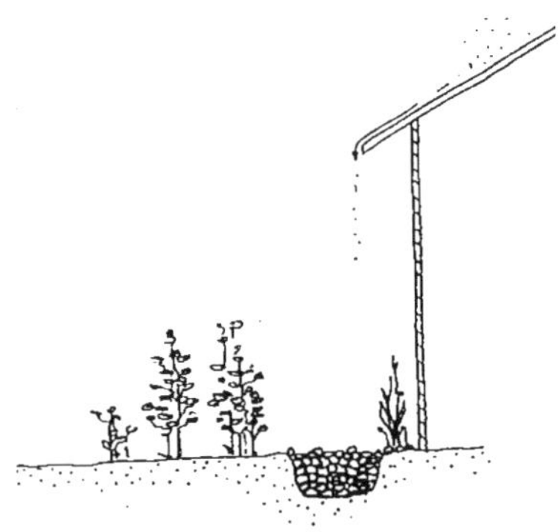

Greywater Reuse at Candlelight Farm

This is a brief description of the greywater disposal and reuse system, which consists of a settling tank and a drain system in an infiltration area.

Typical examples of the determinations and calculations for the required length of drain based on soil type, loading infiltration rate and size of house (number of bedrooms) are given below to illustrate to people what is required before approval for this system is granted.

Method of determining absorptive capacity of soil

Dig a hole 300 mm square and 400 mm deep. Remove all loose material such as roots and stones. Carefully scarify the sides and base of the hole to provide a natural surface into which water may percolate. Add 50 mm blue metal aggregate to protect the bottom from scouring. Fill the hole with water and allow it to soak away. Record the time for the water level to drop at least 25 mm. Use the following table to classify your soil type.

Time for drainage (minutes	Soil type	Loading infiltration rate L/m^2 per day
less than 5	sand	30
5 - 60	loams or gravels	20
more than 60	impervious clays	* see below

* This soil type is unsuitable for the proposed drain system. Other systems such as evapotranspiration systems may be more appropriate.

Note: While the drain area will rely on absorption, it should also be planted out with small shrubs and trees to increase the disposal by transpiration.

Calculations of drain size

To calculate the total length of drain required for greywater disposal you will need to know how many bedrooms the house will/does contain, the loading infiltration rate (LIR) and the proposed drain width and height.

The following calculations are used as an example of a four bedroom house and a LIR = 20.

- trenches 400 (D) x 300 (W) mm
- 1500 mm spacing between trenches.
- house contains four bedrooms @ 240 L each = 960 L (greywater volume only)
- loading infiltration rate = 20 L/m^2.
- surface area of drain required $= \frac{960}{20} = 48$ m
- taking surface of 300 x 300 drain = 0.9 m/lineal length
  length of drain = 48/09 = 53 m

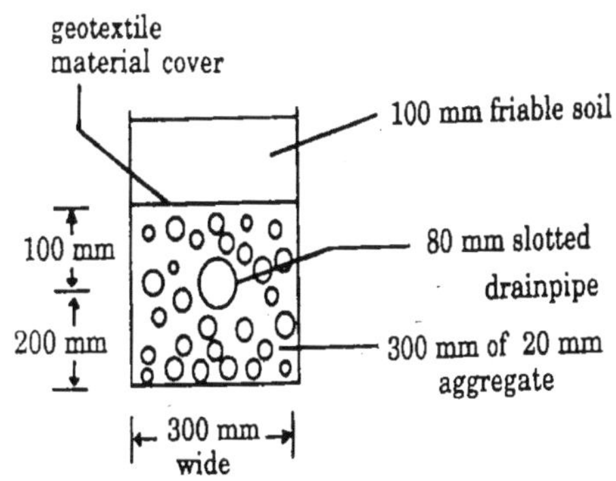

Figure 1. Typical construction and cross-section of drain.

- drain to contain 300 mm of 20 mm aggregate, capped with 100 - 150 mm sand or friable soil with a geotextile membrane between the two. Commercial 80 to 100 mm subsoil pipe will be placed in the upper portion of the aggregate along the complete length of each drain.

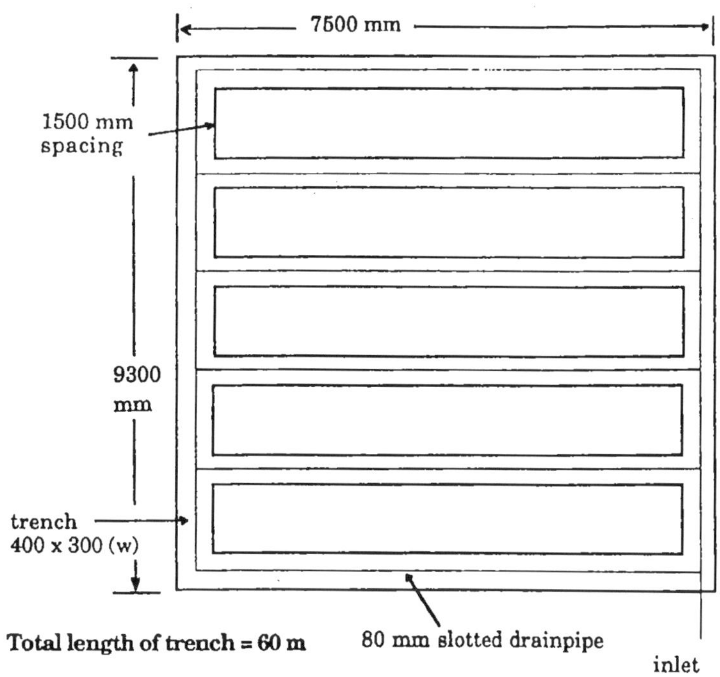

Figure 2. Typical layout of greywater disposal area.

Drain infiltration area

The size of the disposal area and the length of drains mainly depends on soil type. In the example discussed above, if sand were present in the disposal area the loading infiltration rate could be 30 L/m^2 per day. This would mean that the length of drain required would be less.

Furthermore, if the drains were placed closer together the actual disposal area would be reduced. It is possible to place drains with spacing from one to two metres. The closer the drains the less total surface area required.

The diagram below is a typical disposal area layout showing the size of drains and the arrangement of these drains to achieve the recommended total drain length for disposal.

Settling tank

All greywater should initially pass into a settling tank before being pumped or gravity fed into the disposal area.

The Clivus Multrum Composting Toilet

One of the principles of both permaculture and natural systems is that waste materials are recycled on site. Two wastes that rarely get recycled on site are human excrement (solids) and human wastewater (greywater). The non-recycling of these two elements can cause losses of organic matter and large amounts of water from the system.

Large volumes of water are used to transport human manure and urine, dumping them in the nearest river, lake or ocean, turning a valuable resource into an environmental pollutant, wealth into waste. Degradation of natural systems, on which we all depend for our continual existence as a species, is a daily event. All over the world it is happening and it's time it was stopped.

In permaculture systems, water is treated with respect and the reverence that it is due. Water has duties and functions. The objective of the permaculture designer is to enable water to perform the duties and realise its functions to the maximum, before it leaves the sustainable permaculture system it entered.

There are many ways to recycle human and household wastes. Animals, large and small, can be used to recycle kitchen wastes. Greywater, after passing through filters, can be used to water the garden.

Great savings of time, money and energy can take place when wastes are recycled where they originate. One method of recycling human wastes is the use of composting toilets. There are a number of types around. Composting toilet systems collect human excrement and transform them, after a period of time, into humus suitable for placing in potplants or in the garden.

Clivus Multrum is one of the more common composting toilets. It is Swedish, and the name translates into "inclined composting room." Being a completely self-contained recycling system there are very few locations in which it cannot be used. It can be used from the burning deserts to the freezing snow fields.

The Clivus works by positioning a tank of double-thickness insulated fibreglass, containing starter organic material, below the toilet. One or two toilet seats are connected via large diameter pipes to the tanks, and human wastes fall into the tank. A 30° slope at the bottom of the tank causes a gradual creep of the organic material and a constricted space at the front of the tank controls its movement.

The transformed 'old' material is pushed up into the receiving chamber where it is removed at regular intervals as humus.

If operated according to the manufacturer's instructions, it is in a constant aerobic state, that is, oxygen is constantly moving though the waste matter. This ensures high temperatures (60°C) which kills any pathogens (disease-causing organisms), no unpleasant smell, and rapid recycling of organic wastes.

Kitchen and other household wastes can also be placed down a chute or into the toilet directly. This helps maintain a balanced carbon-nitrogen ratio. After a considerable period of time, the manure and urine is turned into rich "garden" compost, which is removed and placed into garden beds or pot plants. This miracle is performed with no power input or moving parts.

The value of the compost product is equal to compost or soil improvement mediums bought at garden centres. There is an important reduction of water use when the Clivus "compost" is applied to the soil. The reduction in the use and consumption of natural resources should be everyone's concern and business.

Little time and effort is needed to maintain the Clivus Multrum, but these essential points need to be observed for a trouble-free run:

- The toilet seat needs to be closed unless in use. This prevents insects entering the system.
- The fan in the air vent pipe needs to be checked and its insect screen kept clear. Both these jobs can be done at the same time.
- The airways leading from the unloading chamber need to be kept clear. The removal of compost from the unloading chamber will keep the airways open.
- The correct carbon-nitrogen ratio needs to be maintained. The addition of other household waste materials such as shredded paper, dry leaves, sawdust and other carbon-based materials will do this.

The little attention the Clivus needs will not take up too much time, and no expensive materials are needed. In fact, owners can become quite attached to their Clivus, and enjoy the brief, regular inspections. The garden certainly appreciates the regular supply of humus.

The cost, however, is more than a septic tank system. However, since it has practically no running or maintenance costs, and lasts a lifetime, a Clivus will, in the long run, be no more expensive than other systems. The production of humus is a bonus which is difficult to put a price on. What price would you place on a healthy productive garden?

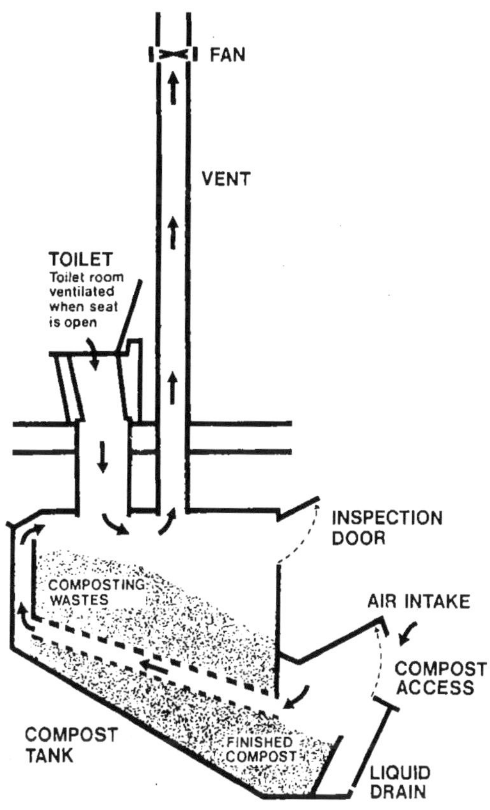

The Dowmus Composting Toilet

The Dowmus Composting Toilet is designed to accept other domestic organic waste, such as kitchen scraps and paper and cardboard, as well as human waste.

The Dowmus is a large enclosed tank which is made from recycled plastic. There's no flush, so plumbing isn't needed.

On installation, the Dowmus is partially filled with active compost, including tiger and red worms. This compost converts the introduced material into more compost and worm castings.

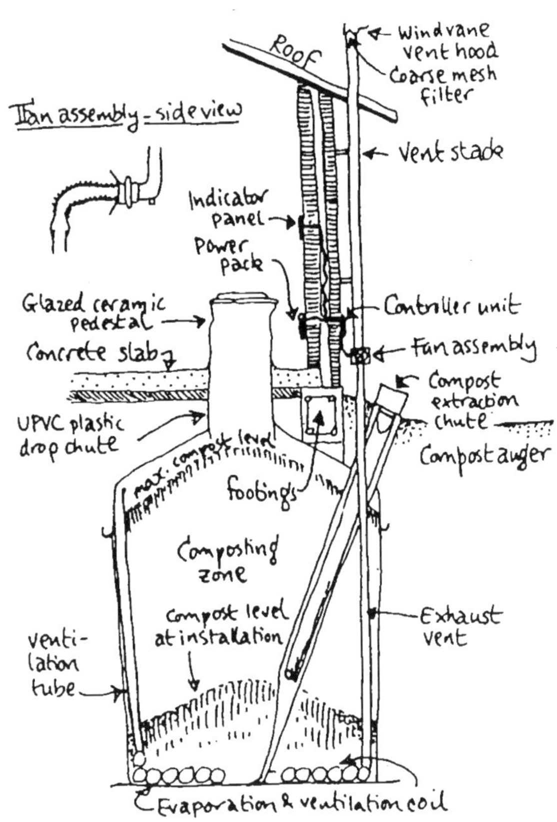

The toilet does not smell because a 12 volt fan runs continuously and draws air in through the pedestal, aerating the compost as it passes through the chamber. Moist air that hasn't already evaporated is vented out through the external flue by the fan.

One unique feature is that the Dowmus can be buried under the house slab. The tank is positioned before the footings and slab are poured. Alternatively, it can be installed above ground or partially buried, depending on the space underneath the house.

The standard unit is approximately 1.8 m by 2 m and the standard kit includes a 240 to 12 volt transformer, a spare fan, a glazed ceramic pedestal with a wooden toilet seat, the compost chamber and all necessary components. The fan can also run from a solar power supply for isolated areas.

A standard unit serving a household of six people should produce about 80 to 100 kg of compost a year. To start with, it may take two or three years for the oldest compost to be well decomposed. When ready, the compost is extracted through a tube with a hand-operated auger.

The auger is drawn out of the tube and the compost simply tapped into a bucket or wheelbarrow - you'll get 10 to 20 L on each 'empty'.

Tanks are fitted to receive all of your greywater as well, so that all wastewater can pass through the system.

Urine in the Garden, or 'Wee Will Fertilise'

Maintaining an intensive urban food garden requires inputs of fertiliser if you don't want to be mining the soil. Sources of nitrogen generally used in city gardening are imported over vast distances and are either petrochemical-based, made from dead animals or exhausted soils elsewhere.

What we tend to forget is that every day people are flushing away millions of litres of their own 'wastes' down the toilet, which could be used as prime nitrogenous fertiliser. Of the 40% of household water that is used in the toilet, the majority is wasted on urine. Up to 20 litres a time of precious fresh water to flush away a cup full of relatively inoffensive wee! Water conservation is another good reason to stop this ludicrous waste, and assume greater environmental responsibility in the process.

It is estimated that the average adult produces enough nitrogen in his/her urine to fertilise approximately $300m^2$ of garden at the annual rate of 70 kg of nitrogen per acre (0.4 ha). This represents both a financial saving, compared to buying nitrogen in, and an important environmental cost saving. As well, urine contains a high proportion of potassium, plus virtually all the other minerals that the body discards. This makes urine particularly good for fertilising onions, peppers, potatoes, celery and carrots. (Ammonia and salts in urine are not well tolerated, however, by such plants as beans and tomatoes.)

Using urine as liquid fertiliser

Composting toilets are becoming more popular these days, but they represent a huge financial investment if built to meet council standards, not counting the fact that you must buy the house first! Another disadvantage with them is that they must be kept dry.

Urine, however, is easily collected if you live in a city. Try keeping a special watering can or lidded container as a urinal, filled four fifths with water, with a lid to stop smells - which indicate ammonia is going to waste. Having several containers is a good idea, as some will have a chance to dry and air in the sun, reducing odours.

When full you'll have a 5:1 mix that the garden will relish when applied to the ground around your plants. Avoid splashing it on leaves if you want to nibble them soon after.

For delicate seedlings a 10:1 water to urine solution can be used. No need to apply urine just before harvest. Fertilising the garden using fresh urine could thus become a daily, evening ritual.

Urine in compost

Another way to use urine is as a compost activator. If you have access to regular supplies of sawdust, a wee bucket is easy to use and doesn't need emptying daily. When it appears damp and before it starts to get on the nose, it's time to tip it onto the fresh compost heap.

Beware of urine leaking into your aged heaps if worms are present, they will be very unhappy. The composting process ensures that any potential pathogens, such as those from the urine of people with hepatitis or bladder infections, are broken down and safe to use in the garden. When fertilising root crops, urine is best applied in the form of compost, to minimise contact between the urine and food.

Urine as weed suppressor

Urine applied to the leaves of most plants may scorch and kill them, due to the high salt and acid levels, so using undiluted urine on hot sunny days effectively kills off weeds. You might need a couple of applications to ensure success.

So in conclusion, if you want a healthy garden that's also environmentally responsible - then using the urine of healthy people is a wonderful way of cycling nutrients, and of getting more in touch with nature.

Information for Schools Wishing to Implement Permaculture-style Gardens

General information

Permaculture uses the simple designs and principles in nature and applies them in garden areas. This becomes evident in the shapes of the garden beds, the importance given to water in the environment and areas where recycling of organic matter occurs.

Permaculture-designed gardens are a very useful teaching tool. They only become effective when teachers use them as part of the curriculum and integrate lessons with hands-on garden activities, such as building, planting, mulching, monitoring, harvesting, composting and recycling. These types of gardens quickly fall into disrepair when they are neglected.

Permaculture gardens become an important link with the larger environmental areas outside of the school grounds. Students can keep small plants and animals in ponds and pens and build mini ecosystems - all of which reflect the kinds of things which happen in the larger, natural ecosystems.

There is a rapidly growing number of primary and secondary schools, both public and private, that are developing permaculture-designed gardens and incorporating permaculture principles in their curricula, after teachers and/or other school staff have undertaken a relevant permaculture course.

At other times, individual parents or community members can be involved in designing and building garden areas in the school.

Costs for these services, if any, are negotiated between the school and the person concerned. These people must be qualified and hold a Permaculture Design Course Certificate.

Conditions for Teaching Permaculture

The word Permaculture can be used by anybody adhering to the ethics and principles expressed in the various books written by Mollison and others. The only restriction on use is that of teaching; only graduates of the Permaculture Design Course can teach permaculture and they adhere to agreed-on curricula developed by the College of Graduates of the Institutes of Permaculture.

Commitment by the school

While the written design and physical work during the implementation of the garden area by parents and other community members may be free of charge, there has to be some commitment to the maintenance and future development of the site by the school and school staff. In particular, the following must usually occur:

1. Involvement and discussion with the Principal, key staff and/or gardener about the plans and concepts for the area.
2. Availability of an area which can be set up to be maintained, with supplies of water and other materials, such as mulch, compost and future plantings.
3. At least one member of the school staff to undertake a Permaculture Design Course - either as part of the design process for the school area or at some earlier stage.
4. Funding for plants and materials for part of the implementation. Nominally, this is a few hundred dollars which usually includes mulch, compost, trees and shrubs, irrigation, and borders and paths.

Usually, some materials can be expected to be donated by parents, course members and commercial establishments. Some organic material is often available on site - from lawnmowing the school grounds or compost and mulch obtained or made by the school gardener.

A continuous supply of organic matter is crucial to the future development and success of the garden areas.

Other considerations

During the implementation stage of the site only some parts of the garden area may initially be built. The important role of ownership in the gardens and the need for students to be involved in the planting and setting-up stages cannot be over-emphasised. This may mean inviting students on a weekend to help build the initial gardens or removing them from class for that day.

Students have to be responsible, with teacher direction, for maintaining the garden beds in tasks such as weeding, picking off slugs and snails and removing dead leaves for the compost heap. They need to be given areas for their own individual or group garden beds and they should feel involved in the development of the site. It is envisaged that the school and staff continue to develop and maintain, with the help of the children, the whole area of the design over the years that follow.

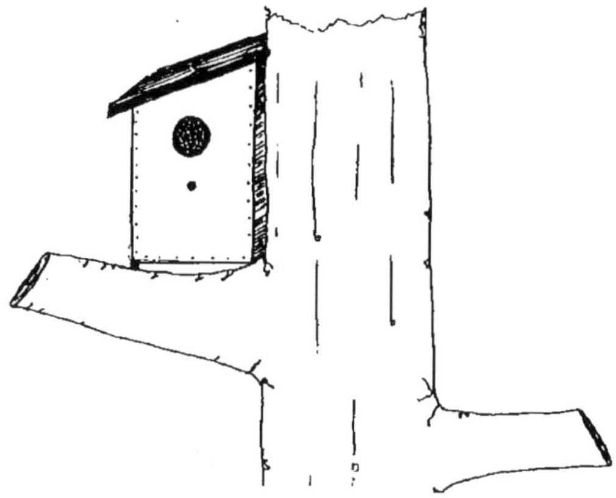

Support

Some support by local permaculture practitioners and any of the Permaculture Associations and groups throughout the world can be expected upon request to them. Contact your local group and find out how they can help.

School Gardens: Grounds for Learning

As we become more globally aware and understand issues such as the greenhouse effect and sustainable use of resources, we need to create opportunities for children to develop positive attitudes towards the wider environment.

Principals are well-placed to encourage staff and students to use school grounds to participate in action for the care and conservation of the environment. A variety of gardening activities can be integrated with existing programs of elementary science and injected with the powerful motivator of meaningful involvement as students see and use the fruits of their work.

Vegetable gardens offer wonderful scope for children to understand and develop skills in soil management and the effects of tilling, weeding, mulching and fertilising. They can see directly the effects of the rhythm of the seasons; of rainfall and sunlight; pest management; and growth requirements of different crops.

They can experience the tingling joy that comes from eating food they have grown themselves from tiny seeds and realise more powerfully than by a thousand words that the future of their world can truly be in their hands - that their effort is significant.

Herb gardens offer similar benefits as produce is easily grown. The folklore and history of herbs is a rich treasure for students to discover, as is herbal use in cooking, cosmetics and medicines.

Children can use some herbs for craft activities such as making pot-pourri or lavender bags and learn the skill of natural insect control through companion planting.

Shadehouse planting of seedlings for local use or re-sale is well-established in many schools. Students can be involved in all stages of soil preparation and growth, and learn the benefits of fruit trees bearing generous crops and shade trees to control the micro-climate and improve soil conditions. Many country schools may already tend flourishing vegetable gardens, but even inner-city students can enjoy these same activities although space may be limited.

Small disused corners, hanging baskets, window boxes and tyre columns can be used to grow salad greens, cherry tomatoes, capsicums, herbs and bright flowers in the city! Try a worm box where thousands of busy creatures tirelessly turn sand and scraps into rich compost for your plants.

Schools can also provide children with opportunities to use their senses, including feeling and reflecting, to discover the environment. These types of programs are a critical factor in leading students to develop a sensitivity to the whole environment and concern for the well-being of planet Earth.

Some schools immerse children in sensory experiences in natural bushland to achieve the same goals. Students bury themselves in leaves in the forest and quietly observe the trees and animals as they have never before.

Recent developments in permaculture offer schools more opportunities to integrate environmental education and elementary science with maximum student involvement.

Permaculture is derived from permanent and agriculture. It is the design and management of agricultural systems which involve the use of plants, animals and humans to achieve maximum production in the least possible area. It is based on three main ethics: care of the earth; care of people; and contribution of surplus time, money and energy to help others. The major departure from normal agriculture is the diversity of plants and animals, each complementing the other, and having several functions.

Low maintenance is essential. Mulching of weeds with newspaper is common, rather than removal by digging. Manures and compost from chickens, worms and food scraps are recycled to improve soil quality and stimulate plant growth, and plant and animal species are chosen to suit unique local conditions.

Permaculture offers our children the chance to understand and develop skills for a system of sustainable agriculture which encourages household and community self reliance.

Its Australian creator, Bill Mollison, notes that suburban lawns use more resources than any other agricultural industry in the world.

In Australian cities nearly 70% of water supplies are used on suburban lawns. He says: "A house with two cars, a dog and a lawn use more resources and energy than a village of 2000 Africans. There is no other path for us than that of co-operative productivity and community responsibility."

These values fit well with our religious views of respect for creation and co-operative efforts to assist those in need. Perhaps we can help children to discover God in the garden. And the exercise and fresh air will be of great benefit to children, while they help to green Australia.

The Outside Classroom - Permaculture in Schools

Introduction

What school hasn't got large areas of under-utilised land? What schools have fully developed their school grounds as a landscape for children?

This article examines ways that teachers can develop an outside classroom, an area in the school grounds that can be used to enhance all aspects of school life.

The outside classroom allows students to learn about water and nutrient cycles, plants and animals, soils, ecology, food production, biological pest control, horticulture, aquaculture and much, much more - right on nature's doorstep!

The main emphasis is permaculture, which is a holistic approach to environmental education. All subject teachers can use the outside classroom, and it is easy to see how classes of english, mathematics, science, health education, art, society and environment and other subjects would benefit.

It must be remembered that not all learning occurs in a classroom. Probably only three-quarters of what children learn actually occurs in the formal teaching process. Students also learn about life, their environment and social interactions and development out of the class - most often in the playground. Teachers need to recognise and acknowledge this and develop strategies to enhance the learning processes which occur in schools.

What is in it for you?

The quality of the school environment has a great deal of influence on a child's behaviour. Unfortunately, many schools do not have abundant gardens, but only barren expanses of bitumen and sparse lawn areas between school buildings. Young people experiencing such an alien environment are unlikely to have any sense of pride in their school surroundings. Thankfully these bleak areas are disappearing as schools take the opportunity to transform their surroundings.

Many educators now realise that creativity is often stifled in the classroom. How can we expect students to write about their life experience or to produce creative writing if they only sit at a desk? If we want students to become imaginative, then we need to provide the ideal learning surroundings.

Developing an outside classroom and making better use of the school grounds can be the inspiration for a whole range of activities in the educational process. It is in the school grounds that the seasons leave their unmistakable signature and it is here too that children create their most uninhibited work. It is said that the best writing is borne out of observation and experience.

The gardens of the outside classroom can be used for many cross-curricular activities. Most subjects, such as english, maths, science, society and environment, and art, can all focus on a similar theme, using the garden as the link. For example, if the science class was studying pond life, students could be writing a story for english about some aspect of the animals that frequent the pond, art students can draw the flowers, frogs or birds around the pond, maths students could measure the depth and diameter (or length and width) to determine the surface area and volume of the pond, home economics students would be investigating the use of pond plants in cooking or salads and so on. Table 1 briefly illustrates some examples of the sorts of activities that can be performed in the garden areas.

Gardening, then, can be thought of as the common ground which links many aspects of environmental education. It enables children to gain knowledge, understand processes, develop skills in observation, communication, recording, problem-solving and sensori-motor manipulation, and cultivate attitudes of care, awareness and tolerance.

Building gardens is not the only activity or the only option when designing an outside classroom.

Any improvement to the school grounds can be seen as useful. In particular, some of the benefits of improving school grounds would include:

(a) offering a greater range and options of teaching strategies which will improve student learning outcomes.

(b) enhancing the quality of the environment, with beneficial changes to the soil, vegetation and the general appearance of the area.

(c) positive improvements in students' behaviour with a reduction in accidents and vandalism, and an increased responsibility towards care and ownership.

Improving the school grounds clearly has considerable educational and social benefits. Students become closer to nature, water saving can occur by better designed gardens and use of mulch in the garden beds, and the micro-climate of the area will change - garden areas are much cooler and more inviting than bare sand areas.

Table 1. Some suggestions for activities in an outside classroom.

Subject	Activities
Science	Leaf litter creatures; pond life; wildflowers and wildlife in the garden; structures - buildings and materials; soil
English	Storytelling; poetry, storywriting and drama - role play
Mathematics	Number (frequency) of slaters, snails, tadpoles, bees and water beetles for pie chart or histogram graph; surface area of leaves; using a trundle wheel to measure garden areas; patterns in nature
Manual arts	Making wooden bird and bat boxes, ceramic bird baths, wire (design and mesh chicken tractors, concreting paths, building compost technology) holding structures and earthworm farms
Society and Environment	Land-use patterns, endangered animals and plants, conservation strategies

Guiding principles

School children usually spend at least one hour each day out of the classroom - in the play areas of the school. I wonder what they think about their surroundings. Can you imagine what they must be thinking about the attitude of their teachers and the school? How often are the teachers just as unhappy as the students about the condition of the school grounds?

How simple would it be for a school to shape the contours, hollows and banks of the ground's surface to create all the micro-climate and aspects necessary for environmental diversity? How simple would it be for schools to convert some of the asphalt jungle into productive garden ecosystems, even to plant trees to create a "green" landscape in a largely urban environment?

We should consider the school grounds as an asset for the whole community and not just the sole property of the school. Too often, large tracts of land are not being used efficiently.

It does not take much effort to re-think the use of school grounds, especially if you can involve local people in concepts such as community gardens.

When building a garden we aim to develop an edible landscape - a healthy, balanced environment in which the plants, soil and wildlife can be nurtured without using artificial chemical feeds, pesticides and herbicides.

In this context, some teachers will have to learn to be land managers and schools will have to manage the school grounds in new and more imaginative ways. Indeed, the most successful development of sections of the school grounds depend on the principles of:

(a) a holistic approach - a design which integrates humans with the landscape and considers all aspects of the curriculum, and the development of long-term goals for the use of such areas.

(b) a participatory effort - the design and development process should involve representatives of the whole school community. This includes the principal, teachers, children, parents, community members and school gardeners.

(c) sustainability - the development should be such that it preserves, and hopefully enhances, the ecology, diversity, productivity and resource base of the area. For example, we encourage teachers and schools to use organic gardening techniques. These are environmentally beneficial to living creatures and the methods themselves are inexpensive and, along with permaculture design principles, provide classic examples of sustainable development.

The main thrust of permaculture is the design. While a design can be represented visually by a 2D map or drawing, as a landscape plan, this is inadequate.

Maps and plans do not indicate the inter-connectedness between all of the components, such as structures, plants and animals we use in the design.

In essence, a design is the marrying and blending of what grows best in the particular area and soil with what the children, teachers and school want.

The design should aim to meet the educational, emotional and social needs of the students. It should also reflect the values, beliefs and cultures of the student population.

While Aboriginal students could plant and maintain bush tucker and bush medicine plants and Asian students could plant and maintain a variety of Asian cooking vegetables and herbs, it may be better to get all students to learn more about these aspects of other people's culture, as well as their own at the same time.

Finally, involve children as much as possible, and particularly in the planning sessions, inviting their ideas and welcoming their suggestions. Although children (and many teachers) usually have little experience about growing plants, creating ecosystems and designing gardens, you will be pleasantly surprised by the suggestions that students make.

Conclusions

Field work has an important part to play in both primary and secondary schools. It provides opportunities for drawing on the environment as a stimulus for learning, and at the same time developing awareness and curiosity about the environment.

We should not under-estimate the importance of play and out-of-class activities in a child's education. We need to acknowledge that there is a distinction between the formal curriculum (what is taught) and the hidden curriculum (what is learnt from the child's surroundings). Furthermore, schools have a responsibility to inform people about environmental issues, and if teachers are comfortable about taking their classes outdoors, then student awareness of these issues often occurs.

The greatest potential for environmental education, which will produce the maximum benefit in student learning, will occur when schools consider changes to the use and management of their school grounds. This will involve both the physical design of the under-utilised areas and changes to the pedagogy, or teaching styles, in schools.

Permaculture Education for Children

Feeding chickens, collecting eggs, tending earthworm farms, and transforming the existing lawns into edible gardens is all part of student life at the Bridgetown Camp School.

The camp school had the grounds redesigned early in 1991. The design had to complement the existing organisational structure of the camp, where students are given the responsibility of completing daily duties. It also had to provide opportunities for students to be involved in "hands on - minds on" environmental education activities. Consequently, students now have the opportunity of converting unproductive lawn areas into a highly productive food succession trail.

This trail takes the form of interlocking hexagon shaped "mandala" gardens. What we call our "honeycomb" gardens. A six-pen chicken rotation vegetable garden, and two worm farms provide additional chores for the children.

The process of designing and constructing the mandala gardens gives the children the opportunity of applying their understandings of the natural world into a man-made edible landscape. Other learning should occur by simply being exposed to the workings of this more diverse, resilient, and stable system: learning through "osmosis", if you like.

Earth Education (which is the process of helping people live more harmoniously with the natural world) activities in nearby forests usually precede the mandala garden activity. This engagement with a healthy and thriving natural system is designed to develop deeper understandings of such ecological concepts as energy flow, cycles, diversity, community, interrelationships, change, and adaptation. What we learn from the bush is later applied to the mandala design and construction.

In addition to Permaculture Education and Earth Education, the Bridgetown Camp School also supports "normal school" curriculum.

The idea of getting students out of the classroom and learning through direct observation and experience forms the basis of the philosophy here.

Farm, forest, historical, art, craft and outdoor adventure excursions are all offered. The potential educational opportunities available here may never be realised if teachers don't develop new ways to educate children.

Teach Your Children Well

When at first I started thinking about the application of a permaculture vision to a school environment, I attempted to identify the problems involved. After looking at various examples in nearby schools, I noticed a theme in all of these examples. Each project was confined to a small area, centred around a mandala garden or tyre pond. For a while these intensively managed areas flourish but rarely do they survive the test of time. Often the teacher involved in establishing and managing these gardens moves on and the driving force behind the garden is lost. The end result is a weed-infested eyesore that attracts negative comments and thoughts about permaculture among teachers and parents.

It would be easy to approach the new school with the idea of having a small permaculture plot in the school grounds, possibly fenced off from vandals and dogs. I feel sure, that with the right approach, the school would even be quite enthusiastic about the idea, but I believe that something new is needed. For example, how can a school possibly promote healthy eating to children in an unproductive landscape where the only food is in the canteen. The way to teach children is by letting them immerse themselves in the experience. To watch a piece of fruit form and gradually ripen, before it is picked and eaten, is a valuable lesson - one that cannot be learned from a book or blackboard.

There are educational benefits to be gained from planting bird-attracting, native plants to provide habitat for insects and reptiles. Again, these habitats will give experiences that cannot be gained inside the classroom.

So we begin to see how the school grounds should look. A mix of fruiting trees and shrubs, nut trees, local native plants, bird-attracting flora, shade trees and interesting areas where children can interact with and manipulate their environment with minimal supervision.

Indeed, the development could be carried out through community-led workshops and courses which would lead to some level of community ownership that may last longer than the time their children are at the school.

Only by offering a simple, low-maintenance, productive hybrid system can the framework for a more extensive permaculture system be introduced to all schools. This system is based on the zoning principle that is used in permaculture design with several notable omissions. Zone 0 relates to the activities within each classroom, and although permaculture has a lot to offer in this area I feel that this is a separate issue. Zone 1 is omitted because of its annual nature and high maintenance (I would describe Zone 1 as an option but not essential to the hybrid system). Other aspects of permaculture design, such as sector planning and plant stacking, are taken for granted in this article but may need to be explained in a submission.

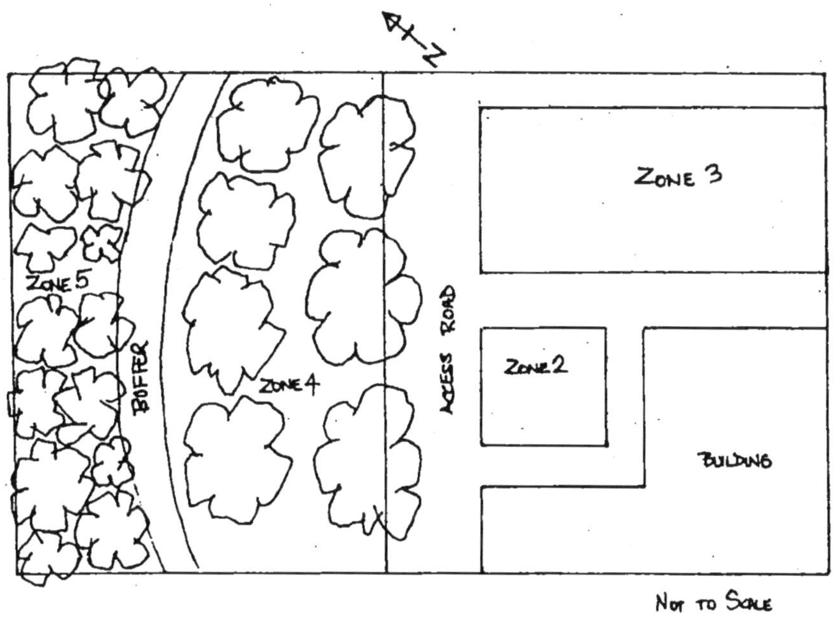

FIGURE 1: CONCEPTUAL ZONE DESIGN

Zone 2 - This is the area within small enclosed courtyards or north-facing heat traps. Nearly all schools have an abundance of these due to the large numbers of buildings and their seemingly haphazard layout. A feature of these areas is absence of lawn and the dominance of paving or concrete to cope with the foot traffic that passes through these busy areas. Ideally these microclimates should be well reticulated, heavily mulched (possibly with a living mulch), and fenced in some way. Some species that may be useful in Zone 2 are banana, tamarillo, guavas, feijoa, sugarcane, pawpaw, selected citrus, sapote (white and black), blueberry, Japanese Raisin Tree, brazil cherry, fig, passionfruit, pepino, pomegranate, sweet potato, rosemary, strawberries and comfrey.

Zone 3 - In larger grassed courtyards and long sheltered avenues between buildings, mixed orchards of larger trees can be grown to provide fruit, nuts, shade and other useful products. These areas are usually covered by reticulated lawn that will not be adversely affected by a mix of well-spaced evergreen and deciduous trees. Some suggested species for Zone 3 are olive, paulownia, date palm, custard apple, mango, jackfruit, avocado, almond, hazelnut, pistachio, macadamia, Irish strawberry tree, capuli cherry, icecream bean tree, tipuana tipu, apple, pear and loquat.

Zone 4 - The open areas away from buildings can be used for growing larger trees with an emphasis on valuable timber species and other tree crops. Zone 4 will also be reticulated to the lawn beneath but the trees will be hardier and lower maintenance than Zone 3.

Trees will be pruned high to discourage climbing and to increase the value of the timber. This zone will cover the highest percentage of the school grounds and if planted correctly would create a substantial asset for the future in the form of high quality cabinet timber and perennial tree crops.

Species selection includes carob, oak (such as holm and turkey), pecan, chestnut, walnut, macadamia, tamarind, olive, tipuana tipu and stone pine.

Buffer - Between Zones 4 and 5 is a densely planted wall of shrubs and small trees that act as a fire retardant buffer. This buffer also functions as a living fence to minimise foot traffic into Zone 5 or to channel it towards appropriate paths into Zone 5.

The buffer can also be used in narrow areas to screen classrooms from busy roads to reduce air and noise pollution from the road. Special attention should be paid to the density of planting and selection of species that will shade out any grass or weed growth that may become a fire risk or impact on Zone 5.

Useful species for the buffer include *Coprosma repens*, pigface, Natal plum, agave, sea grape, pomegranate and tagasaste.

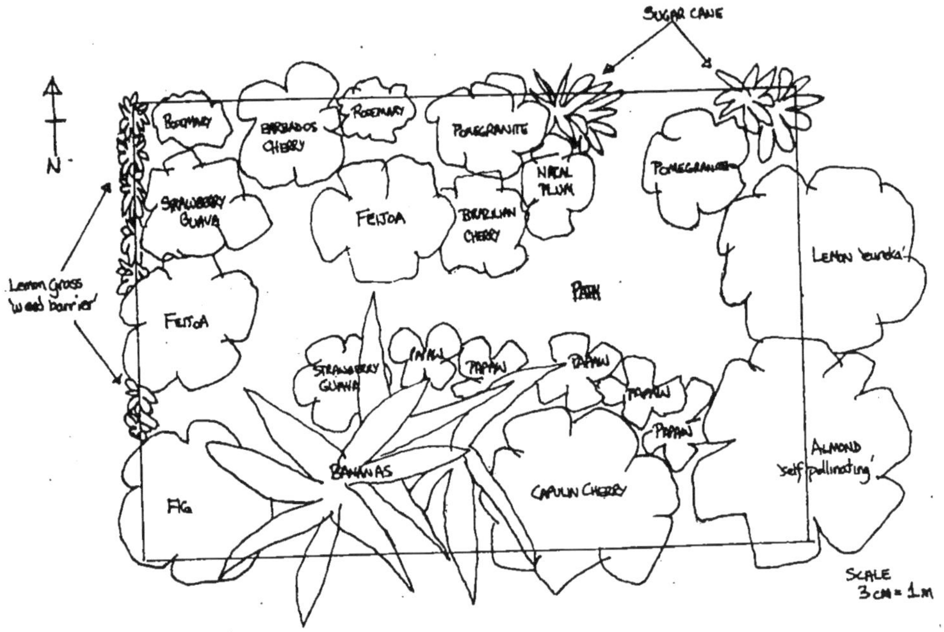

FIGURE 2: SAFETY BAY SENIOR HIGH, ZONE 2 CONCEPT 'DREAMING'.

Zone 5 - Zone 5 is usually an area of wilderness or bush. The areas furthest from the school buildings will be given over to bush regeneration if this is appropriate and should be kept as close as possible to the original state of the bush in the region.

Pathways through Zone 5 will be kept to a minimum to reduce the risk of weed invasion; also Zone 5 should be as wide as possible to avoid the negative impact of the edge effect. The numerous local and national groups that have a special interest in bush care and regeneration may help to fund bush regeneration at the school.

In conclusion, I believe that with support from the school and some funds, a permaculture-designed system will produce a pleasant, productive, interactive, interesting and environmentally-sound school grounds.

Permaculture Kalumburu Style

The school garden

It was evident, not only from my conversation with the school principal, but from my own observations and discussions with Aboriginal community residents, that tremendous improvements have been made in the school.

Of particular interest was what I observed behind the school's home economics building. In what had once been a seldom-used section of the school's lawn, a thriving permaculture garden could now be found. Was permaculture partly responsible for the improved relationship between the school and the community? Did the school's permaculture program contribute to the improved attitudes of the community residents and the sudden appearance of home gardens?

The garden teaches the children much more than just gardening. The children are learning very valuable skills, such as co-operation and planning. They are learning that what they put into a project is what they get out of it. They are learning about recycling, composting, and the environment.

One thing that is important is that the children are the "boss" of the garden. It is *their* garden and they make the rules for it. They take responsibility for it. It is seen by them as their garden. Since it is their garden, they care about it and look after it. They are also less likely to vandalise it.

Each child has their own plot or area to take care of. Often this area is shared by two other children. The children are responsible for that plot, watering it, caring for it, and deciding what to plant there.

"We have an interesting barter system for compost. The compost for their plots is not free. The children must buy compost by purchasing it with other material. The children are really getting into it. They pick up garbage from around the school and community, collect old boxes and cardboard from the community shop, and pick up leaves around the school yard. The material is then divided up into its appropriate bin. The children can purchase compost for their plot, learn about recycling, and keep the school yard and community clean at the same time. The children must work hard for the compost."

There is something else that is really important. The students, of all ages, are taking the knowledge they have gained in the garden back out into the community.

They are talking to their parents about what they've learned in the school garden. But, more than that, they are implementing it, trying it out at home. That's really great to see. They are communicating with their parents and starting their own home gardens.

Parents and community members are starting to come into the school to see the progress of the garden. What's important is that the children are showing their parents around the garden and explaining to their parents and grandparents, the concepts of the garden, the projects they are working on, and the function of the garden. What's even better is that the parents really seem to be listening. Just look around the community. Look at the houses and what do you see? Cleaner houses, less garbage, and neatly kept gardens.

The high school students have constructed a shade house. They have a small greenhouse for cuttings, seedlings, and young plants. It's working very well.

The children have recently constructed tyre ponds in order to attract frogs: the frogs have become their pets. The children are also learning about good and bad insects. There are plenty of good insects, such as ladybugs and earthworms in the garden.

Harmful insects tend to proliferate within the school's permaculture garden. Citrus scale and cucumber beetle seem to be the worst offenders. However, the effect of harmful insects tends to be offset by the numerous beneficial insects which also proliferate within the garden. There is no shortage, for example, of frogs, earthworms, ladybirds, and spiders. The positive and negative effects of insects on the permaculture garden is a topic of study and discussion amongst the Kalumburu school children.

Positive perceptions

The positive effects of the school's permaculture garden on the Kalumburu Community should not be underestimated. The permaculture garden was introduced to the Kalumburu school as a student-directed initiative in 1995. Since that time, the garden and its positive ramifications, which have spread throughout the community, have continued to grow and flourish.

The Kalumburu example clearly illustrates that one is never too old or too young, or too isolated or too far gone to learn from, benefit, and appreciate the positive effects and perceptions gained through permaculture.

One year later

The changes to the permaculture garden were incredible. The children of Kalumburu School have continued to expand and develop both their knowledge of permaculture and the garden itself. Where there was once an isolated and disused section of the school lawn, a lush and thriving permaculture garden now stands.

Not only the plants are thriving in the hot north Kimberley conditions, but also the school children, who are enthusiastically participating in the permaculture program.

The school day now consists of an interesting, exciting, and fun-filled dimension that is working and learning within the permaculture garden. This 'hands-on', practical and enjoyable initiative has all of the students talking.

Permatours

Tourists to our school are able to have a permatour - a guided tour of the permaculture garden. Four children are involved in any one tour. Two of them will race off and start making lime juices for the tourists or pick some food from the garden, like a tomato or some goose berries, depending upon what is in season. The other two children will take the tourists on a tour of the garden and answer all of their questions.

The permaculture tours give the school children a real purpose for practising their 'city talk' (ie. standard English).

They answer the tourists questions using the very best of their 'city talking'. All of the children in Kalumburu need to know two languages. They have their own 'home language', but they also need to know how to speak to the people in the towns and the cities.

By conducting the permaculture tours with the tourists, the school children are also leaning the kadia ('white-fellas') way of manners. For example, they learn to say 'you're welcome', 'yes please', and to wash their hands before eating. A lot of these manners are quite alien to these children. They just haven't been taught them before. So, this gives their education a real-life basis.

Garden update

The Kalumburu School children have incorporated an increased number of plants and additional permaculture concepts, such as rotational chicken vegetable yards/pens, into the garden.

The chicken pen was originally full of weeds. Now there are no weeds at all. The soil is extremely healthy, fully organic, full of worms, and very soft and moist. All the school children will do is rake it up. We will plant some vegetables. However, most of the yard will be left to crop on its own. That is the idea of permaculture; that it will look after itself.

We also have some vegetables growing in pots with some signs. The school children made the signs themselves, in library studies. So, the other classes are supporting the garden. Some unruly children came into the garden during the school holidays and used the signs as missiles. So, many of them got ruined.

The sugar cane and bush tucker plants have taken right off. We just can't eat them fast enough. The Aboriginal family who lives next to the permaculture garden is very happy because all of the pumpkins go over there to fruit. The pineapples, which were donated by the mission, are also fruiting.

Enough food is produced in the permaculture garden to feed every child in the school once a week with a big plate of food. There is usually so much food produced in this garden that most of it goes back to the chickens.

A way forward

The isolated Drysdale River Mission settlement (now Kalumburu) has come a long way since a handful of Benedictine priests sank the first metal spade into the far north Kimberley soil in 1908. The plants which would annually refuse to grow for the early missionaries are now thriving within the school and home gardens. The mission gardens, expansive and set apart from the rest, continue to grow.

However, through permaculture, both the school and the community have adopted a new interest and concern for their community and their environment.

Successful Method of Raising Trees from Hard-Shelled Seeds

Preparing the seed

I have tried two methods of preparing the seed and both have been successful.

One method is to scarify the seed (that is, to file through the hard outer coating of the seed opposite the area where the seed will sprout.)

The other method is to place the seed into water that has just gone off the boil. The water volume is approximately 20 times the volume of the seed. Allow the water to cool down with the seeds in it and leave for 12 to 24 hours. Observe the seed size before and after the soaking and you will see that some of the seeds will have swollen. These are ready to plant. Repeat the hot water treatment with the remaining, unswollen seeds.

The germination method

Materials required

1. Two old newspapers 12 mm thick.
2. Four litres of well-composted soil, worm castings, or a mixture of both.
3. Water.

Preparation

Set up your newspaper nursery in a site protected from wind, in a sunny position in winter or in a shady spot during summer.

1. Lay one newspaper on the ground and soak with water.
2. Place 40 mm soil in the centre of the paper, leaving a 75 mm border of no soil around the perimeter of the newspaper. Water well.
3. Do not let the seeds dry out. Place them approximately 25 mm apart on the soil surface.
4. Cover seeds with 10 mm of the remaining soil and water well.
5. Place the other newspaper over the soil and soak with water.

Management and maintenance

1. Keep the newspaper continually damp.
2. After one week, lift the top paper and look for germinated seeds (the seeds of the carob, tree lucerne and leucaena will germinate in one week if the conditions are right).
3. One week after germination, carefully remove the seedling and place into a pot. When lifting a seedling of any type, use forefinger and thumb on the leaf only. Do not hold by the stem. Always keep the pot moist.
4. You can transplant your seedlings after 1 - 4 weeks. I have found that immediately after the first week is best, as there is less root growth to be damaged.

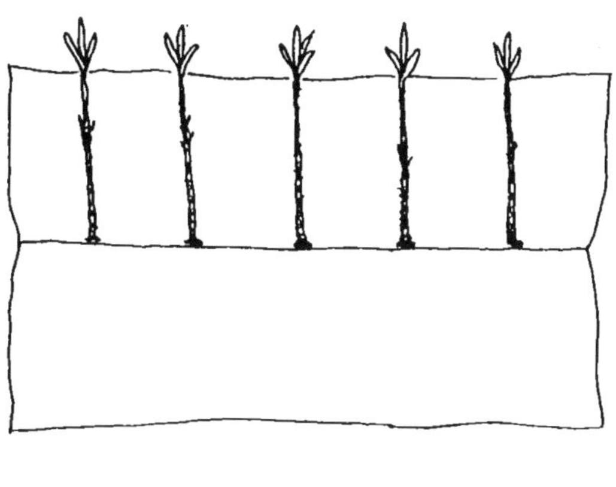

Growing Native Tree from Seed

November is a good month to start growing trees from seed. An early start means a longer growing season, producing well-grown trees large enough to plant out the following winter.

If you have ever grown vegetable or herb seedlings, you can grow trees. It is just as easy!

Start with good quality seed - either collect it yourself or buy a commercial packet.

1. Use a good seed-raising mix: half sand and half well-made compost or worm castings.
2. Collect trays or seed boxes at least 50 mm deep - a foam fruit tray from the green grocer is a good size. Fill trays with soil.
3. Sprinkle seed lightly on top, cover with a thin layer of vermiculite or seed-raising mix. For large seeds, such as Marri, bury the seed to about its size.

   Germination success rates vary; some will be patchy and others cover the tray quite thickly.
4. Put under a partially shady tree or shadecloth until they germinate. Keep well watered, but not too wet.

   You can cover the trays with a sheet of glass or clear plastic if nights are cool, but remove these when the seeds germinate. Damping-off is the problem to avoid here, so check them regularly.
5. When they have developed their first 'true' leaves start pricking them out. Use a pointed popsicle stick, plastic cocktail fork or something similar, and lift them carefully out of the tray.

   Place the seedlings in small pots or tubes, filled with the same soil mix. You can add a little fertiliser to each pot.

   Make sure that the plant roots are straight; if coiled they might tend to continue to grow this way.
6. Firm plants down, water in, and place back under the shady tree. When they are growing well, place them in full sun, but don't let them dry out too much.
7. They will need watering at least once a day during hot weather. Many people go on holiday at this time, so keep this in mind. You may need to find a reliable neighbour.
8. Weekly liquid feeding with fish emulsion, or applying blood and bone or a slow release fertiliser is essential. Your seedlings need to be healthy if they are to compete with weeds and survive in tough conditions. Generally, ease off fertilising and watering in winter.
9. Plant out with the early winter rains, from late April or May onwards. They will not make much top growth in winter; rather, they are busy putting down roots.

   The plants should 'take off' in spring. Protect them from rabbits, kangaroos and other herbivores. Keep weeds at bay with an application of mulch around the seedling.

How to Take Cuttings

What you need:

- Plants. If you are taking your cuttings some distance from where you will pot them up, you will need to keep the cutting ends moist, so placing them in a plastic bag or cup of water is a good idea.
- Peat and coarse sand or river sand or a propagation mixture.
- Either plastic pots or a deep seed tray.
- Water.

There are three types of cuttings that you can take - tip cuttings, cuttings taken with a heel and hardwood cuttings.

A heel cutting is taken from a side branch of the main plant which is pulled off with a downwards movement so that you finish up with a little piece of the older wood attached to it. This is known as the 'heel' as that is what it resembles. At this point you should remove any flowers or seed heads from your cutting as we want the plant energy to go into forming new roots rather than into flowers or seed production. Also remove all leaves except for a few top ones.

With the aid of a pencil (or popstick) make a hole in the propagation mix and firmly plant in your cutting. Remember that cuttings do not require any compost or fertiliser at this point. Keep cuttings moist until roots form. This is a very popular method used for propagating plants such as lavender, santolina, sage, rosemary and geraniums during either the spring or autumn.

A hardwood cutting is a larger piece of wood, usually taken from a tree such as an elderberry taken during late autumn and left in the open ground over winter. It usually sprouts leaves and roots the following spring. The cutting needs to be taken from mature wood and it is cut straight across at the base where the leaves join the stem and cut at an angle (so that any water can run off) above another pair of leaves.

A tip cutting can be taken year-round apart from winter and is the easiest way to propagate herbs. A tip cutting is taken off the end of the young growing stem of a plant. A piece between 3 and 8 cm will suffice. Try to ensure that you have at least 3 gaps in between leaf joins. Cut straight across below where a leaf joins the stem. Remove the bottom two pairs of leaves. Pot up into the propagation mixture, placing the cuttings in just below the next leaves. Cuttings can be placed quite close together all around the edge of the pot. Keep moist and reasonably shaded in a warm sheltered spot.

You will find in a few weeks that the cuttings will have formed new roots. You can use a hormone rooting powder if you wish; however, I tend to never have any on hand when I need it, so I often don't use anything at all and have still had reasonable success.

Honey has often been touted as having similar properties to some hormone powders, otherwise if you happen to have a willow tree in your backyard, cut off a few branches, break them up and soak them in a bucket of water. The resulting water will have a hormone-stimulating effect on plant cuttings of various sorts.

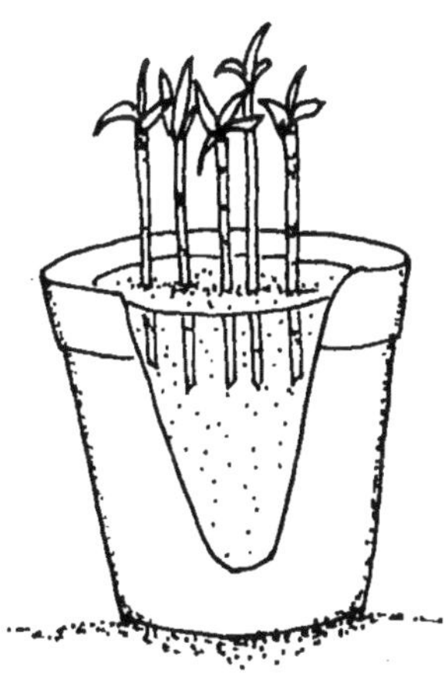

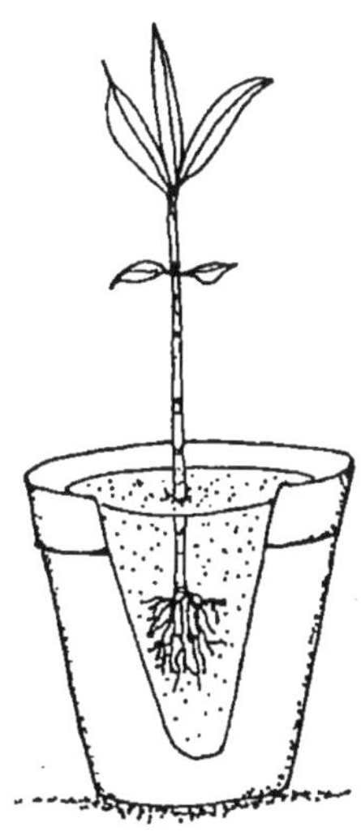

Once cuttings have grown new roots and shoots (and you will easily recognise this by the fresh green appearance of the tip growth), it is time to lift them carefully, disturbing the soil surrounding the roots as little as possible.

Either plant into the garden or into individual pots. This is now an appropriate time to give your new plants a feed of compost or fertiliser. Water them in and keep shaded for a few days before you harden them off in full sunlight.

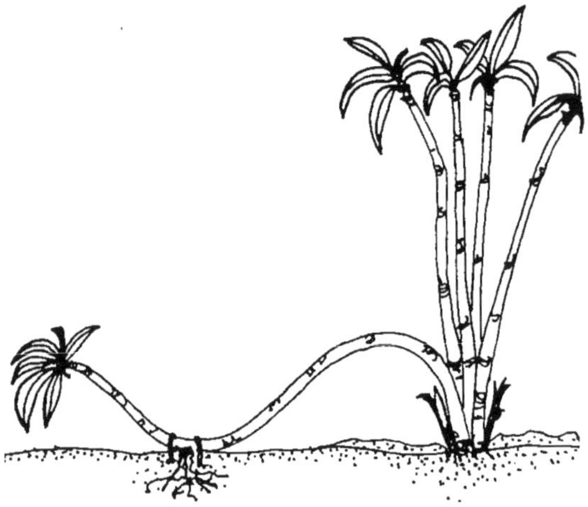

Tip cuttings are a suitable means of propagating herbs (but not umbelliferae or annuals and biennials) including mints, sages, balm, bergamot, basils, thyme, hyssop, geraniums, winter savory, sweet marjoram, tarragon, wormwood, southernwood, horehound and catnip.

Bay and lemon verbena tip cuttings can be taken during summer when there is fresh new growth on the trees.

Herbs which die down in winter (tarragon, bergamot) should be taken in spring or early summer.

Other methods of propagating herbs

These include root division and layering. Layering consists of pegging down or earthing up a stem, so that it is partially under the soil, where it will form roots. This is then cut away from the mother plant once new roots have been formed. Try this with mints, sage and thymes.

Root division is dividing up a plant by severing the root between two growing portions of plant. Try this with yarrows, mints, tansy, some artemisias, lemon balm, orris, marjoram and bergamot. Use this method in spring or autumn.

Vegetable Seed Saving

Botany

There have been various systems of plant classification through the ages in an attempt to prevent confusion between countries and different names for the same plant. Linnaeus, in the seventeenth century, devised a system of classification using Latin names, which were based on the flowering characteristics of each plant. There have been difficulties with his system as more plants have been discovered and named.

The newer system considers plant modes of nutrition, structural features and reproduction. All names are still in Latin, and each plant has one Latin name that cannot be used for any other plant, thus ending the confusion between countries and cultures about exactly which plant is being discussed.

Linnaeus' system of classification is still being used. It begins with kingdom (whether plant or animal) then phylum, class, order, family, genus, species and variety.

Family: this is where most botanical classifications begin in books and catalogues. This is the really important end of the scale for saving seeds.

Genus: this is the name always written with a capital letter. Plants in the same genus generally will not cross-pollinate.

Species: this is the name always written in the lower case and plants of the same species will cross-pollinate. It is important to understand and remember how this part of the Latin naming system works.

Variety: this is not necessarily a Latin name. It is a variety within a species and will cross with any other plant from the same species. The variety name should be indicated as such.

Life cycle of plants

The largest group of plants on the Earth are the flowering plants (Angiosperms). How long the plant lives and how often it flowers are the next points to know.

There are three types of plant life-cycle:

Annual - when the plant grows, flowers, reproduces and dies within one year.

Biennial - when the plant grows and produces an edible crop in the first year, then goes to seed in the second year and dies.

Perennial - when the plant usually lives two or more years, producing flowers and crop each year.

Types of flowers

All flowers are sexual: that is, they have male or female parts, or both. Flowers that have both male and female parts in the same flower are known as "complete" flowers; those having only one part are called "incomplete" flowers.

Complete flowers may be self-fertile and self-pollinating, such as peas and beans. These are the easiest types of plants to save seeds from. Complete flowers may also be self-sterile: that is, they must have pollen from another plant to pollinate. This is called cross-pollination. Cross-pollinated plants tend to be genetically stronger and more adaptable, and therefore survive better in a natural ecological system. Vegetables which cross-pollinate from different flowers on the same plant include carrots and beetroot, whilst many of the cabbage family cross-pollinate from flowers on different plants.

With incomplete flowers, both male and female flowers may appear on the same plant (for example, corn and the pumpkin family) or they may be on separate plants (for example, asparagus).

Agents of cross-pollination

Insects are the most common means of cross-pollination. Bees are essential for relatively few plants, whilst wind serves as an agent for plants with light pollen.

Sorting seeds

Seeds are living organisms: they need to breathe. Do not keep them in plastic or sealed tins and jars. Unsealed tins, jars, paper, hessian or calico bags are all suitable. Seeds can also be stored in the freezer, but, in this case, need to be well sealed in plastic.

Heat and humidity shorten the shelf-life of seeds. The worst place to store them is in the garden shed. Usually, garden sheds fluctuate between the extremes of temperature and humidity. The best place is a cool, dry, dark place such as a cupboard, pantry or similar area.

Freezer storage of seeds extends their shelf-life to double or longer. It is the way they are kept in the seed gene banks in various parts of the world. It is also a useful way to kill weevil eggs under the seed coat of beans, peas and corn without the use of chemicals.

Germination testing

I use the damp paper towel method. I only use about 10 to 20 seeds: I'm not after great accuracy to the last decimal point, just whether some, most, nearly all or nearly none are viable.

Some common vegetable seeds

Peas and beans (family Leguminosae)

Most peas and beans are annuals, but a few beans are short-lived perennials (Scarlet Runner and Hyacinth for up to 7 years). A few beans do cross within the species, such as Broad beans and Scarlet Runner beans. For saving, keep varieties 3 m apart, preferably with a tall crop between (corn etc). Varieties that do cross (Broad beans) should be isolated - 250 m apart. There are two ways to save seeds:

1. Mark off a section of row and save every pod in that section. Remove any weak or off-type plants. Leave the plants until most of the leaves are dry, then pull them out and pick off the pods. Dry the pods further if necessary then thresh and winnow, or keep the pod whole.
2. Mark individual pods on bushes with a metal twist tie. Pods can be picked when turning brown, or when seeds rattle inside the pod.

Do NOT save just the residues and remnants of the crop when you pull it out; it will encourage traits like lateness. Also, educate the rest of your family about what is being saved for seed and how it is marked.

Seal and freeze the seeds for at least 24 hours in the freezer to kill any weevil eggs under the seed coat. Pea and bean seeds last two to three years.

Lettuce (family Compositae)

All lettuces are the species *Lactuca sativa* although there are four types of lettuce - crisp-heading or hearting, loose-heading or hearting, cos and leaf. There are also dozens of varieties of each type of lettuce. Separate varieties by three metres for saving.

Lettuces will cross with wild lettuce and with Celtuce, which is an ancestor of lettuce. The cross is easy to pick; the resulting plants are very bitter! Temperature seems to be the triggering factor in sending lettuces to seed. Select two or three of the best, earliest, tastiest (if a leaf or cos), most solid heart, and allow to go to seed. Most lettuces go to seed in early summer, though more heat-tolerant types bolt in late summer. Do not save seed from "bolters"; that is, plants that run straight up to seed without forming decent hearts and leaves.

The seeds look like thistle seeds. They ripen unevenly so it is generally necessary to pick off the early ripe fluffy ones once or twice, then when most are white and fluffy, cut the whole head. Hang the heads in paper bags to dry further. Separating the fluff from the seeds is a pain! I use a combination of rubbing and winnowing, a very messy and time-consuming job.

Lettuce seeds keep for four to five years.

Tomatoes (family Solanaceae)

Tomatoes are self-pollinating and crossing is virtually nil if they are spaced over two metres apart. However, don't grow them next to each other if you are saving them for seed. Allow selected fruits (mark with a twist-tie) to get very ripe but not rotten before picking.

It is important to ferment tomato seeds so as to destroy seed-borne diseases. To do so, scrape out the pulp and seeds into a glass or ceramic container (not metal, although plastic may be alright).

Add a bit of water and allow it to ferment for two or three days, stirring occasionally. Rinse the seed clean under a tap (I use a metal strainer) then spread them out to dry on paper towels.

Tomatoes are grown as annuals but can be kept as perennials.

Capsicums and Chillies (family Solanaceae)

Capsicums and chillies are botanically the same species (*Capsicum frutescens*) and are generally regarded as annuals, although they are perennial in tropical areas. I have had plants go through winter, although frost kills them.

They are self-pollinating, although bees will cross them as well and you should isolate different varieties by at least three metres. Preferably isolate individual flowers or groups of flowers if you have several varieties by wrapping them in old nylon stockings or something similar.

Simply keep the seed when you pick and use the fruit. When handling chillies be extremely careful because the seeds are very "hot" even when replanting. The seeds keep for about four years.

Eggplant (family Solanaceae)

Scrape out seeds of very ripe fruit, dry then store. The plant is an annual but will crop for several years, with a very long growing season, and is self-pollinated, although insects will cross it if more than one variety is being grown. Seeds are viable for a couple of years.

Onions and Leeks (family Liliacae)

Both onions and leeks are insect pollinated. Leeks can be left in the ground, where they will flower in the second year of growth. Hence selection is simply a matter of leaving some of the best stems in the ground until they flower.

Seeds do not fall out of the seed-head when ripe so there is no problem about losing any, just pick when dry.

The seeds are hard to get out of the seed-head and must be threshed or rubbed then winnowed. Onions are picked when the tops are dry then stored in a dry airy place. When the bulbs shoot again, replant and they will flower then seed.

Select the best shaped bulbs that are slower to shoot. Onion seed falls out of the seed-head when ripe, so when you see some black seed, cover the head with a paper bag and let it continue to ripen. When the whole seed-head is dry you can cut it and the seed should thresh out easily. (Onions can also be left in the ground as for leeks).

Onions cross with other onions but not with leeks. You can have many varieties growing to eat, but only one at a time when flowering. The seed is short-lived and only keeps for up to two years.

Beetroot and Silverbeet (family Chenopodiaceae)

They cross-pollinate with each other and are wind pollinated (very light pollen), so keep them isolated as much as possible. They are biennial and beetroot may need frost (chilling requirement) to flower and seed.

Cut stalks when two-thirds of the seeds are brown and partially dry then thresh after two or three days. If left longer the stalks dry out and are harder to separate out. The stalks need cutting several times and the seed needs further drying after separating. The seeds remain viable for up to seven years.

Spinach (family Chenopodiaceae)

Spinach is annual and may have four kinds of flower types in the same variety of plants: separate male and female flowers on the same plant, only female, only male or no flowers at all. The flowers are small and inconspicuous and are wind pollinated. Spinach bolts to seed as the temperature rises and days lengthen. The seeds mature as the plant dries. To harvest, pull out the plants and strip the seeds by hand.

It is extremely difficult to maintain purity unless the plants are isolated or all of your neighbours grow the same variety - the pollen drifts for kilometres.

Seeds last for five years.

Carrots, Parsnip, Celery and *Celeriac* (family Umbelliferae)

Parsnips and carrots don't cross whilst celery and celeriac are botanically the same and do cross. All of them are insect pollinated.

For carrots and parsnips, allow large roots to remain in the ground. In the second year they will send up a flower stalk. Pull out plants that go to seed early or do not form good-sized roots first.

With celery, only allow large well-formed plants that have produced lots of stalks to go to seed. All tend to ripen unevenly so several cuttings of ripe seedheads are usually required.

Parsnip seed only keeps one year, carrot seed three years and celery seed should keep for five years.

How to Store Seed

Vegetable and other types of seeds are at their peak for storage when they reach maximum dry weight on the plant. The most vigorous seeds at harvest time will keep the longest in storage, provided they are stored correctly.

Drying

High seed moisture is the greatest single cause of loss of viability and vigour in storage (vigour is the seed's ability to germinate and grow). Exceptions are the fleshy seeds of certain tropical fruits, palms and most citrus (called 'recalcitrant seeds'), which can be stored in a cool moist situation without drying.

Fresh harvested seed of high moisture content will often generate heat if left in bulk, so seed should be spread out in a thin layer in the sun to dry.

However, the higher the seed moisture, the lower the drying temperature ought to be. Rapid drying using excessive heated air can be harmful, as rapid shrinkage of the outer parts of the seed can shorten seed life and impair germination. In general, drying temperatures should not exceed 42°C.

Silica gel crystals can be bought and used to absorb moisture from seeds in a sealed container. The colour of the crystals indicates how much moisture has been absorbed - blue if dry and pink for moist. Use 1 kg silica gel for 10 kg seed and packets. Re-activate the silica gel at 120°C in an oven until the beads are deep blue. They can be dried an indefinite number of times.

As long as the seeds are kept dry in moisture-proof containers, they can be stored even in very humid, hot climates. Seeds are best stored in envelopes or paper bags which are then placed in a dark jar or tin (with a tight-fitting lid to keep out moisture).

Temperature and humidity of the storage environment

Temperature and relative humidity are the most important factors affecting seed quality during storage. Relative humidity, which is the amount of moisture in the air compared to what it can be at that temperature, is the most important of these, as seeds are hygroscopic; i.e. they easily release or absorb moisture from the atmosphere.

Generally, seed storage life is doubled for each 1% decrease in moisture content, while the problems of maintaining seeds in storage increase with seed moisture content. This can be generalised as follows. If the seed moisture is about:

* 40-60% - germination occurs.
* 18-20% - heating may occur.
* 12-14% - fungi grows on and in seed.
* 8-9% - insects become active and reproduce.

Similarly, seed storage life is doubled for each 5°C reduction in storage temperature. Refrigeration to at least 5°C is desirable (this is slightly warmer than domestic fridges). The cooler the temperature the more slowly seed vitality declines, even at temperatures below freezing. At 5°C and below, insects also become inactive and this helps to control insect damage to seeds. Thus, high temperature and high moisture will kill seeds rapidly as will invasion of micro-organisms and insects.

Pest management

Before storing any seeds, it is worth making sure that weevils and other insects are not already in residence. The eggs of weevils and other insects hide under the seed coat of some types of seed and emerge when the temperature and moisture are right.

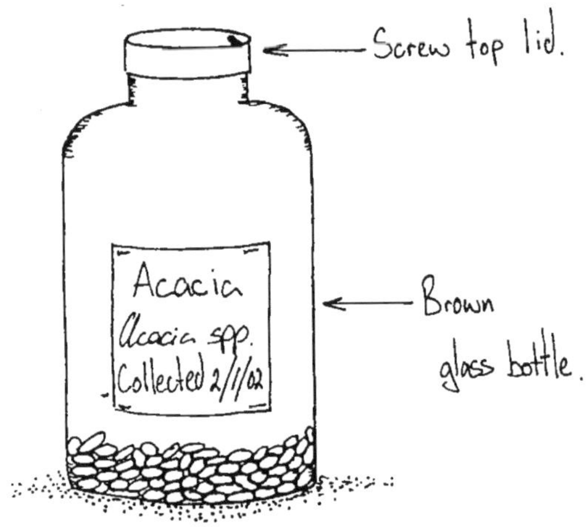

Two days of freezing inside a sealed container kills most weevils and eggs - although some species of insects would need much longer at lower temperatures.

African farmers thinly coat beans and other large seeds with edible oil, renewing the oil if weevils are seen. If seed moisture content is maintained below 8%, and the storage is partially or completely sealed so that respiration reduces the oxygen content below 14% (air is 21% oxygen), insects cannot survive in the seed. Sachets of oxygen absorbers contain, basically, iron filings which absorbs the oxygen - they will go hard when spent.

Seed viability

Viability is how long a seed can live in a dormant state. Eventually, seeds lose their capacity to germinate. Some seeds can last several years, others need to be planted almost as soon as they are harvested.

It is always best to find out how long your particular seed will last and put this information on the seed packet - a sort of 'use by' date.

Place seed packets of similar viability in the same jar or tin and label the front with a list of what's inside and the 'plant before' information.

Good luck. Remember, it is vitally important that each of us harvest and maintain our heritage varieties of herbs, vegetables and fruit trees, and pass these on to others and on to our children.

Handy Hints for Propagation

Propagating seeds with long root systems

For those who haven't already thought of it, a very effective way of propagating seeds, especially those with a long tap root, is to make further use of those old newspapers. Roll sections of newspapers into a tube and then secure the tubes with two to three pieces of twine (see figure below).

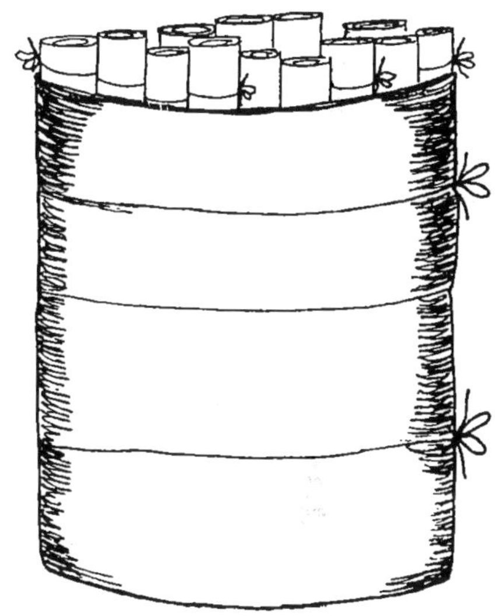

Make up about ten to twelve tubes and place them into a 10 L black plastic planter bag. Use enough tubes to ensure their security, then cut the base from another planter bag and slide it carefully over the remaining exposed area of tube. Secure these with twine and fill the tubes with good earth and plant your seeds.

When it's time to transplant them to their permanent spot, there's no need to disturb the root system, just lower the tube in - taproot and all.

Germinating sandalwood seeds

Carefully cut the hard outer layer with a bandsaw to allow the water to penetrate the kernel. Be careful not to damage the kernel. If only germinating a handful, or if a bandsaw is not available, use a hacksaw. Cut at a tangent to give more control and to leave a flat area. Place the seeds inside a hessian bag and lie out on the ground so they form a sandwich.

Keep moist until germinated. The bag allows for easy inspection. Plant out those which germinate, bearing in mind that they are root parasites and will need other plants around them to survive - but preferably not couch grass. Quandong seed can be treated in a similar way.

Growing cuttings

Materials required

- A strip of polythene film
- Sphagnum or Peat Moss
- Hormone powder or honey
- 8 to 10 cuttings

Place the moss on one half of the plastic strip. Prepare the cuttings by dipping them in the hormone powder or honey then place them on the moss so that the cut end is at the bottom of the moss.

Fold up the bottom of the plastic strip to cover the moss and cuttings then roll into a bundle and tie with some string. Water occasionally. When the cuttings have rooted, plant them out in the usual way. A lot of native plants have been grown successfully by this method.

The cuttings may be half-ripened shoots after seasonal growth or hard wood cuttings from mature plants.

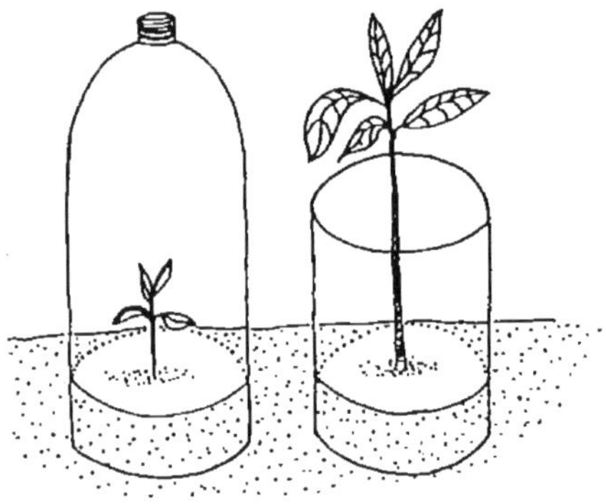

Mini greenhouse

Two litre, clear plastic drink containers make good mini-greenhouses if their bases are removed.

Germinating parsley

Place the parsley seeds on hessian and fold over the hessian to cover the seeds. Pour on boiling water and let the seeds soak.

They will germinate in approximately four days and can be placed into pots.

Transplanting fruit trees

There comes a time when most of us find that we either have a fruit tree in the wrong position for some reason or other, or we have to shift to another residence and wish to take our trees with us.

The following method is the way I have been 100% successful, even with aged lemon trees, moved 15 km and a nectarine moved 300 km.

Firstly, a trench the width of a spade is dug to the depth of 40-50 cm, just outside the dripline. If this is done in winter, then just leave this for a week or two. If in summer, cover the severed roots with hessian and keep damp.

If possible prepare the new site in advance. This is not always possible if shifting from one house to another, so if the move is in summer, dig the hole before starting the next step on the tree to be removed.

When the day arrives for the shift to be made, make sure you have appropriate transport and a hose at each site long enough to reach both tree and proposed site.

Prune top of tree to about one-third to one-half normal size. Remove bags or hessian from around roots, dig trench a little deeper in two sections on either side of tree and start washing the sand from the rest of the roots.

From time to time the trench has to be cleaned out to allow sand to be easily washed from roots. This process is fairly time consuming.

Transport tree with roots covered in plastic or hessian bag for protection, and place in prepared hole and reverse procedure by washing sand/soil into all crevices of roots while tree is held in position. Any damaged roots should be removed prior to this process with a pair of sharp secateurs.

Keep well watered till signs of good growth appear.

Other general hints

- Get all seedlings growing strongly in individual pots before planting out.
- Home-grown bean seeds should be stored in the freezer, or they won't keep.
- Seeds from parsnips are viable for years. Even three year old seed can be successfully germinated.

What a Wasteful Lot or a Permy's Paradise

by Jeremy Parker

One fine day, I was going along, cycling to the shop.
Home again to get the car as I'd seen something tip-top.
The Council's roadside rubbish day was due on the calendar,
"Look at all that useful stuff" I began to say.
Steel and pipe and wire and drums, I'd need some help with this.
I'll save myself a lot of money re-using this alleged rubbish.
So off we went down the street to gather this useful junk,
we'd only been around the block before the suspension went clunk.
The second load was a little slower but the collection an assortment.
"Where are we going to put this stuff, you'll need a bigger allotment."
Station wagon and eight by five, some people looked aloof,
"There go those bloody Permies with a cane chair on the roof."
The following day, I must say, could have been a little boring,
except the bug had hit me so hard I had to do more exploring.
"The fridge! The fridge! It's gone, Oh what rotten luck.
I'll have to get a B class license and next time hire a truck."
"The new area by the golf course, we haven't been down there yet."
So off we went into the sun to what proved a good bet.
Bikes and tyres and wood and hoses. A wood bbq, hardly used!
An aviary with a little rust, not to be abused.
Another load from farther away.
A wave and smile from a guy in a bus, no need to worry about him
'cause he was doing the same as us.
So now we stare in despair at this pile of accumulations,
"Think of all the money we've saved." Hmmm, exaltations!
So if you have some useful junk you must really throw away,
leave it out for the Council truck and make a Permy's day.

Another e-book in this series

Permaculture Volume 1: The Best of PAWA

Selected articles from the Newsletters of the Permaculture Association of Western Australia. 1978 to 1990

First published in 1996 and featured articles on the themes of:

- Permaculture basics
- Useful plants in permaculture
- Fruit trees
- The uses of trees in rural permaculture
- Basic techniques
- Aquaculture
- Poultry and other birds
- Other animals in permaculture

Made in the USA
Monee, IL
13 September 2026

62981008R10077